Astronomy through the Ages

Astronomy through the Ages

The story of the human attempt to understand the Universe

Robert Wilson

Not from the stars do I my judgement pluck,
And yet methinks I have Astronomy
Shakespeare

CRC Press
Taylor & Francis Group
Boca Raton London New York

CRC Press is an imprint of the
Taylor & Francis Group, an **informa** business
A TAYLOR & FRANCIS BOOK

CRC Press
Taylor & Francis Group
6000 Broken Sound Parkway NW, Suite 300
Boca Raton, FL 33487-2742

First issued in paperback 2019

© 1997 by Taylor & Francis Group, LLC
CRC Press is an imprint of Taylor & Francis Group, an Informa business

No claim to original U.S. Government works

ISBN-13: 978-0-7484-0748-4 (hbk)
ISBN-13: 978-0-367-40088-0 (pbk)

British Library Cataloguing-in-Publication Data
A CIP catalogue record for this book is available from the British Library.

Library of Congress Cataloging-in-Publication Data are available

Visit the Taylor & Francis Web site at
http://www.taylorandfrancis.com

and the CRC Press Web site at
http://www.crcpress.com

Dedicated to the memory of a Durham miner

Full many a gem of purest ray serene
The dark unfathom'd caves of ocean bear:
Full many a flower is born to blush unseen,
And waste its sweetness on the desert air.
Gray

Contents

List of Illustrations

The development of life on Earth (p. 6).

The constellation of Orion (p. 17). (Photograph courtesy of Dr David Storey of Worthing, Sussex.)

The zodiac of Dendara (**Plate 1**, colour section).

Star trails (**Plate 2**, colour section). (Photograph with kind permission of Dr David Malin, supplied courtesy of the Anglo-Australian Telescope Board.)

Aristotle's view of the Universe (p. 30).

The measurement of the size of the Earth by Eratosthenes (p. 33).

A partial lunar eclipse (**Plate 3**, colour section). (Photograph courtesy of Duncan Copp and David Heather, University of London Observatory.)

Ptolemy's world model (p. 36).

The world model of Copernicus (p. 56). (By kind permission of The British Library, 59i6.)

The first reflecting telescope, built by Isaac Newton (**Plate 4**, colour section). (By kind permission of The Royal Society.)

The separation of sunlight into its different colours (**Plate 5**, colour section).

Cassini's measurement of the distance to Mars (p. 95).

A snooker game analogy to the second law of thermodynamics (p. 109).

Measurement of velocity by the Doppler effect (p. 112).

The phenomenon of radioactivity (p. 118).

Einstein's principle of equivalence (p. 132).

Quantum models of the hydrogen atom (p. 146).

The distribution of stars in terms of their luminosities and colours (**Plate 6**, colour section).

A selection of spiral, elliptical and irregular galaxies (**Plate** 7, colour section). (Photograph with kind permission of Dr David Malin, supplied courtesy of the Anglo-Australian Telescope Board.)

The relative size, nature and colours of the planets (**Plate** 8, colour section). (By kind permission of Professor Jay M. Pasachoff, taken from his book *Astronomy: from the Earth to the Universe* (1991).)

The orbits of the planets shown to scale (**Plate** 9, colour section). (By kind permission of Professor Jay M. Pasachoff, taken from his book *Astronomy: from the Earth to the Universe* (1991).)

The main parameters of the planets expressed in values relative to the Earth (p. 185).

Composite image of one part of Venus taken by the spacecraft *Magellan* (**Plate** 10, colour section). (Courtesy of the National Aeronautics and Space Administration (NASA).)

An image of the Martian surface taken with the Viking spacecraft (**Plate** 11, colour section). (Courtesy of the National Aeronautics and Space Administration (NASA).)

An image of the local Martian surface taken with the Viking lander (**Plate** 12, colour section). (Courtesy of the National Aeronautics and Space Administration (NASA).)

The *Trifid Nebula* (**Plate** 13, colour section). (Photograph with kind permission of Dr David Malin, supplied courtesy of the Anglo-Australian Telescope Board.)

The core of a solar-type star at its red giant stage (p. 212).

The *Helix Nebula* (**Plate** 14, colour section). (Photograph with kind permission of Dr David Malin, supplied courtesy of the Anglo-Australian Telescope Board.)

The core of a massive star at its red supergiant stage (p. 218).

The 1987 supernova explosion in the Large Magellanic Cloud (**Plate** 15, colour section). (Photograph with kind permission of Dr David Malin, supplied courtesy of the Anglo-Australian Telescope Board.)

The Pleiades star cluster (the Seven Sisters) (**Plate** 16, colour section). (Photograph with kind permission of Dr David Malin, supplied courtesy of the Anglo-Australian Telescope Board and The Royal Observatory, Edinburgh.)

Preface

This book grew out of an undergraduate course that the writer gave at University College London on the foundations of modern astronomy, but with a difference. It was aimed at the non-scientific faculties and therefore had to be non-mathematical; as such it was unique in the UK. The reason behind the course was to broaden the education of students in the humanities, but the book has a much wider purpose and has been written with that in mind.

In 1959, C. P. Snow, an accomplished scientist and distinguished novelist, wrote about what he called the "two cultures", by which he meant those individuals whose training and activities were scientifically and technologically based on the one hand, and those whose training was in the area of the humanities – classics, history, literature, and so on – on the other. His point was that each culture knew little about the other, that communication between the two was difficult (and sometimes impossible) and that this situation was bad for society as a whole. The writer agrees with this view by Snow, which was based on his experience in the 1950s, and believes that the separation is even more marked in the 1990s and even more damaging. There is a very clear absence of scientifically trained personnel in the influential areas of society – in politics, in the civil service, in the media, in business and industry. This observation is not made as a prelude to an argument in favour of more scientists in influential positions (I have met many who would not be up to running a coffee stall), but it is an argument in favour of those in such influential positions knowing *something* about science, because it touches on so many important issues of the day. However, in the writer's experience (confined mainly to the UK), many influential people not only know nothing about science or technology, they are almost proud of it and can hardly wait to proclaim the fact. Such individuals far outnumber those scientists who disdainfully reject an interest or knowledge in the fine arts, who would regard the *David*s of Michelangelo and Donatello as chunks of white and black marble, Rembrandt's *Night Watch* as a splash of paint, and Goya's *Maja* as a tawdry attempt at pornography. It is to their counterparts in the humanities that this book is addressed in an attempt to communicate a knowledge of science, in the form of astronomy, and to illustrate the fascination and beauty of the subject.

Of necessity, the presentation is non-mathematical (even graphs are excluded), but the subject matter is not simplified in any way. The deep concepts of space and time, of relativity and quantum mechanics (the two classic examples of the incomprehensible to the general public), and of the nature and origin of the universe, are described and discussed without resort to the meaningless analogies that so often mark works in popular science. The aim of this book is not to impress but to inform and at the same time, hopefully, to interest and to entertain. Sadly, this aim seems to be the inverse of that of much of the media in presenting scientific matters; this is particularly evidenced, at the very moment this Preface is being written, by the levels of hysteria that were reached in reporting findings of some components of a meteorite from Antarctica, but presumed to have come from Mars, as representing definite evidence that life had existed on that planet.

This is an appropriate moment to mention the scientific method, which the writer hopes will be better understood after the reading of this book. It was first practised by Kepler, then defined and elucidated by Galileo, and finally brilliantly demonstrated by Newton. It has been the basis for the great scientific advances of the past four centuries and is not really a method but a discipline, or a set of rules by which science should be pursued. It states, very simply, that all claims, all propositions, all hypotheses and all theories should be subject to the test of explaining *all* available information, and then be subject to a further test of predicting some new effect which can be investigated and measured. The scientific method is therefore completely opposed to dogma and requires that all claims and statements should be verified by an appeal to the available facts. Some readers may feel that this is so obvious that it does not need saying, but, unfortunately, dogma has played a major role in history, and still does today in many important areas of human activity.

As mentioned earlier, this book developed from an undergraduate course for non-science students. Those students taught the writer a great deal, in particular, that they more easily absorbed science if it was placed in a human context. This determined the structure and theme of the book as being the story of the human endeavour to understand the universe.

The book is written for the intelligent lay-person and is therefore directed at a very wide audience. The one person the book is not intended for is the professional astronomer and, just in case one should read it, I should explain that I am not following the strict scientific practice of referencing all the contributions of individual astronomers, but only when such reference aids the story for the non-scientist. Hence, particularly in the final chapters, several hundred astronomers who have contributed to the developments outlined are not mentioned and the writer makes this broad acknowledgement now.

Although the book is strictly non-mathematical, there may be technical terms used with which the lay-reader is unfamiliar. I have tried to explain these in the text as they arise but, as a backup, I have also compiled a glossary at the end of the book and the reader should consult this whenever necessary.

Finally, I would like to say that I hope this book will project a knowledge and feeling for the fascination of astronomy to a wide audience. Writing it has been a very personal activity, but one that has been shared by one other person who processed the manuscript, commented on its content, suggested many of the quotations used, and was a pillar of support throughout – my wife Fiona. It is our book.

Robert Wilson   August 1996

PART I

THE EARLY DEVELOPMENTS IN ASTRONOMY

This part of the book covers the period when observations of the Universe relied entirely on the unaided human eye with its consequent limits and constraints. It starts in ancient times when the earliest civilizations showed a fascination for astronomy, albeit with astrological overtones, and it ends with the Renaissance when the subject was put on to a firm scientific footing and when a new age was heralded by the invention of the telescope.

CHAPTER ONE

The Beginning

Before the beginning of years
There came to the making of man
Time, with a gift of tears;
Swinburne

About five thousand million years ago, a new star was born in our Milky Way galaxy. It was an average star, neither over-bright nor over-faint, and was formed like all other stars by condensation of the interstellar gas under the all-pervading force of gravity. In all such contractions some degree of rotation exists and this causes the material to form into a circulating disk within which condensations occur to form a number of gravitationally bound objects. Whether these objects become a star or a planet depends entirely on the amount of material condensing in the gravitational contraction. If this is large enough, the interior will heat up to the ultra-high temperature needed to ignite a nuclear furnace which generates enormous energy by fusing the most abundant element, hydrogen, into helium. In our Solar System, only the Sun reached this critical mass and the other condensations resulted in the formation of the planets. In more than half of other star formations, more than one body exceeded the critical mass to become a star. This is demonstrated by the fact that more than half of the stars in the sky are multiple, with two or even three stars in orbit about themselves, possibly with some planets.

The interstellar gas from which the Solar System was formed was composed mainly of hydrogen (74 per cent by mass) with 24 per cent helium and all the other heavier elements from carbon to uranium contributing only 2 per cent by mass. These elemental abundances are reflected in the composition of the Sun and the giant planets, Jupiter, Saturn, Uranus and Neptune, but the Earth and the other terrestrial planets, Mercury, Venus and Mars, are rocky in nature, indicating a chemically selective process in their formation which favoured the heavier elements and allowed most of the hydrogen and helium to escape. Yet the heavier elements did not exist at the start of the Universe when the primeval matter produced in the Big Bang was composed entirely of hydrogen and helium, with minute traces of lithium, beryllium and boron; but there was no carbon, nitrogen, oxygen or any of the other 87 elements found on Earth. Hence, the very first stars that were formed in the rapid star-burst era that marked the beginning of our galaxy contained no heavy elements, and any planets formed at that time could not have been even remotely like the Earth. But many of those early stars were more massive than the Sun and, consequently, evolved more rapidly to the extent that they had gone through their whole life-cycles by the time the Solar System was formed. As will be related later in this book, they had successively fused hydrogen into helium, helium into carbon, nitrogen, oxygen, silicon and all the other elements in the periodic table up to iron; then, in an immensely explosive event called a supernova, all the elements heavier than iron, from cobalt to uranium, were formed, and these, together with the lighter elements, were hurled into space in a high-velocity expanding shell. Far from being a contamination of the primeval interstellar gas, this was, as far as we are concerned, a crucial enrichment of it, because it provided those elements essential to life. Indeed, apart from the hydrogen present in the water of our bodies, all the other elements that constitute more than 90 per cent of what we are made of are the result of nuclear processing in the interiors of massive stars and the cataclysmic explosion that heralds the end of their life. We are children of the stars.

These astronomical processes, the nuclear synthesis of the heavier elements in stellar interiors, their ejection into the interstellar medium, and the selective condensation of those elements in the formation of the four terrestrial planets in our Solar System, set the scene for the great miracle – the development of life on one of them, Earth, whose size and distance from the Sun were just right. But the development of life and its evolution was slow . . . very, very slow.

The Earth was formed four and a half thousand million years ago but almost a full thousand million years were to pass before the first micro-organisms appeared and a further thousand million years before marine algae and primitive plants started to generate pure oxygen, not tied up in carbon dioxide, into the atmosphere until, when the Earth was three thousand million years old, it reached a critically important stage for life and the environment when it had an oxygen-bearing atmosphere similar to that of today (except for artificial pollutants). At this point, evolution accelerated greatly and, over the next thousand million years, fish, insects, toothed birds, large reptiles and primitive mammals appeared. Then, 65 million years ago, a catastrophic event, whose cause has just recently been established as a giant meteor or asteroid, led to the extinction of the dinosaurs. It was then that the mammals proliferated and, some 3–4 million years ago, the first human types emerged. But evolution of our own species, *Homo sapiens*, did not occur until a hundred thousand years ago and the earliest civilizations did not develop until after the most recent ice age had ended about 12000 years ago.

To give some idea of the timescale of the development of life on Earth, the important milestones are listed in the table which also scales real times to one year, that is, as if the Earth were formed on 1 January and its present age is midnight on New Year's eve. This demonstrates vividly the very slow initial development, and then the very rapid later evolution of life, together with the relatively brief presence of *Homo sapiens*. Since this story relates the attempts of the human race to study and understand the Universe it lives in, it is confined, in human terms, to the very brief period of civilization which, on scaling to one year, covers only the last two minutes; but in astronomical terms, it goes back to the very beginning when, most astronomers now believe, the Universe started in a single, immense, explosive event – the Big Bang.

The timescale of the development of life on Earth; in the final column the times are scaled to one year as if the Earth was formed at the start of the year, and now is midnight at the year's end.

	Event	Years ago	As if
	Earth formed	4500 million	1 January
	First micro-organisms	3500 million	23 March
	Marine algae, primitive plants	2500 million	12 June
	Multicellular plants introduce free oxygen into atmosphere		
	Atmosphere like today	1500 million	1 September
	Evolution greatly accelerated		
	Fish, primitive reptiles, insects	300 million	7 December
	Large reptiles, toothed birds, primitive mammals		
	Dinosaurs disappear	65 million	7 December
	Mammals proliferate		
	Human types appear	3–4 million	5 p.m, 31 December
	Homo sapiens	100000	12 minutes to midnight, 31 December
	Civilization	14000	2 minutes to midnight, 31 December
	Modern industrial society	200	1.5 seconds to midnight, 31 December

CHAPTER TWO

Ancient Astronomy

Awake! for Morning in the Bowl of Night
Has flung the Stone that puts the Stars to flight:
Rubaiyat

About 10000 years ago, the most recent ice age was over, the ice had fully retreated and the resulting warm period led to a spread of forests, vegetation, fish and mammals. The human race, which had hitherto spent its energy and ingenuity on survival – the acquisition of food and provision of warmth – responded to the greatly increased food supply by accelerating the development of tools and the techniques of hunting and gathering of plants. The increased productivity allowed the human race to exploit its greatest gift, a powerful brain, more fully. It embarked on the development of civilization and found that the setting up of organized societies, in the form of tribes or whatever, resulted in greater prosperity and more effective defence. It found that farming the land to produce the crops that it wanted, and the husbandry of animals to produce the meat that it needed, were far more bountiful than gathering vegetation and hunting game which happened to be present naturally.

With the development of agriculture and the human control of the environment, the land area needed to support a community decreased by a factor of about a hundred compared to that for hunting and gathering. This great increase in productivity caused an increased growth in population; new societies grew and developed with their own structures and customs to become more unified but also more separate; trade between them developed but, not infrequently, disputes occurred which often resulted in warfare. The victor usually took everything and subjected the conquered to slavery, a practice that became extensive in all ancient civilizations. One powerful form of society that developed was the city-state, which often dominated its surroundings as an empire; Babylon was to be the first and Rome the greatest.

By *circa* 5000 BC, food production had reached a level that freed significant time and effort for pursuits beyond those needed for survival alone. The consequent release of human ingenuity caused an acceleration in human progress, with the further development of new tools and techniques, allowing more effective farming (and warfare), the building of cities, the development of language, from spoken to written, the strengthening of social organization and authority, and the creation of wealth. Out of this evolved the first true civilizations: societies with the means and the wish to pursue intellectual, artistic and other creative activities in addition to the most basic needed for survival. The great early civilizations were located in many parts of the globe, usually in the most fertile regions, often watered by great rivers. The first of these was the Sumerian civilization, which was fully established in Mesopotamia by *circa* 3500 BC, so named because it is the land between two rivers (the Tigris and the Euphrates), and is largely embraced by modern-day Iraq. Others were well established in Egypt by *circa* 3000 BC, in India (the Indus valley) by *circa* 2500 BC, in Crete (Minoan) by *circa* 2000 BC, in China by 1500 BC, and in Central America (the forerunners of the Incas and the Aztecs) by 1000 BC.

Many of the intellectual activities pursued in the early civilizations grew from the innate curiosity of the human race in the natural world in which it existed. This was particularly true of astronomy, where the motion of the Sun, Moon, planets and stars caused excitement and puzzlement. You should realize that life in a modern industrial society is a great impediment to viewing the heavens because of artificial lighting and smog. One of the really beautiful sights in nature is of the clear night sky in the absence of city lights, say on a remote mountain, where the sky has great depth and the Milky Way is a bright lane of light. But this sight was available to everyone in the distant past and its beauty led to the pursuit of astronomical studies in all the early civilizations.

But the early developments in astronomy were fired more by practical and mystical rather than scientific considerations. The development of agriculture required that crops be planted in spring and harvested in autumn; hence the times of the seasons needed to be known. In other words, a calendar was required, and several attempts to establish one were made in the period between 5000 and 1000 BC. Natural time-periods were available in the day, determined by the rising, setting and rising again of the Sun; the month, determined by the time it took the Moon to pass through all its phases; and the year, determined by the seasons, over which the Sun reached its maximum noon altitude in mid-summer (for the Northern Hemisphere), its lowest in mid-winter and back again in mid-summer. The first and simplest astronomical instrument, the gnomon, was able to give some indication of the time of day and the season of the year; it consisted of a straight vertical rod and is based on the same principle as the modern sundial. In the early morning, its shadow would be long and point roughly westwards; as the day progressed, it would shorten and rotate until, at noon, it would be at its shortest and would point exactly due north; it would then lengthen and rotate until, in the late afternoon, it was pointing roughly eastwards. If the length of the shadow is measured at noon, it is found to be shortest in mid-summer and longest in mid-winter, thereby allowing the seasons to be estimated. Another way to tell the time of year was afforded by the night sky. The stars were fixed and unchanging relative to each other, but appeared to rotate completely over one year, so that there were winter and summer constellations.

Another group of objects in the night sky were the five planets or wandering stars. As bright as the brightest stars, they moved in the same plane (the ecliptic) as the Sun and Moon, but in odd and seemingly unpredictable ways. Three of them (Mars, Jupiter and Saturn) would advance across the celestial sphere, reverse their motion and advance again; the other two (Mercury and Venus) also moved but were visible only when they were close to the Sun, just after sunset or just before dawn. The puzzle of the planets was to remain unsolved for several millennia and it posed the greatest problem in early cosmologies – the explanation of the motions of the Sun, Moon, planets and stars.

The mystical aspect of studying the heavens, astrology, developed strongly in the early civilizations and soon became the prime driving force. It was believed that the stars and planets controlled human destiny and therefore their study was encouraged as a means of predicting, or explaining, human triumphs and tragedies. Perhaps this is not surprising: if the heavens could say when crops should be planted or harvested, why not when wars should be embarked upon or preparation made for famine or flood? Religious aspects also crept into the interpretation of the heavens, and astronomical studies were often carried out by priests.

The most ancient civilization, the Sumerian, was based in Mesopotamia, now the southern part of modern-day Iraq. It prospered rapidly and by *circa* 3000 BC had developed a written language which was etched into clay tablets. Unlike the other early civilizations of Egypt and China, it has not retained its name, culture and identity over the centuries, but has changed hands frequently through invasions, migrations and wars. The development of building technology allowed for larger and larger human groupings into city states, and one of these, Ur, became the capital of Sumeria in *circa* 2500 BC. It lay near the junction of the Euphrates and Tigris rivers in the south of Mesopotamia. It is believed to have been the home of Abraham and his semitic tribe, who developed a belief in a single, unseen, all-powerful God, a belief that forms the ancient roots of today's three great monotheistic religions. Abraham was to take his tribe out of Ur *circa* 2100 BC, when it set off on its wanderings (from which the name Hebrew, or "wanderer", derives), which were to take it to Canaan (Palestine), to Egypt and back to Canaan over a thousand years.

In about 2000 BC Ur was conquered by the Elamites, who came from what is now southwest Iran, and a new powerful city-state emerged, Babylon, which was to establish control of Sumeria and extend its empire over the whole of Mesopotamia. Babylon is on the Euphrates about 80 kilometres south of present-day Baghdad. It became very wealthy and very indulgent in the pursuit of pleasure and consumption, giving it a reputation as a magnificent, worldly, wicked city, which still persists today more than two millennia after its demise. Great buildings were constructed, including the Hanging Gardens, one of the seven wonders of the ancient world. The Sumerian cuneiform script was given a syllabic form, thereby greatly increasing its flexibility. Many written tablets still survive which tell us more about the Babylonian Empire than we know about many European countries during the Dark Ages of AD 500–1000.

Many activities were encouraged, one of which was a study of the heavens, entirely for astrological purposes, since they believed their destiny could be read in the stars, but this had the important result of producing the first major set of astronomical observations ever undertaken. Over centuries, the positions of the bright stars were established and the motions of the Sun, Moon and planets charted against the background of those fixed stars. All these data were recorded in cuneiform script etched into clay tablets, many of which are still available today; they allowed some astronomical analysis, as well as astrological interpretations. By about 1000 BC, lunar eclipses could be predicted with reasonable accuracy and the motions of the outer planets had been established over several centuries.

Studies of the heavens also had religious overtones, and the prime celestial bodies were named after gods, of which the Babylonians had several. But in addition to the religious and astrological applications, there was an astronomical requirement – the prediction of the dates of important events such as festivals and the times of sowing and reaping. This required a calendar and one was constructed which had a year of 12 lunar months, and which started with the vernal equinox in spring, the time for sowing. Each month was defined by the Moon and started at sunset on the evening when the new crescent Moon was visible for the first time. Since a lunar month is not an exact number of days, nor is a solar year an exact number of lunar months, this was a somewhat cumbersome basis for a calendar, as many other civilizations discovered. A lunar month is close to 29.5 days, so, with each month being identified by the first appearance of the new Moon, a month was either 29 or 30 days and a year of 12 lunar months consisted of 354 days. The difference between this and the solar year (365.25 days), which determines the seasons, was accommodated with an extra month about every three years. The Babylonians also introduced the seven-day week in deference to the Sun, Moon and five planets, regarded as representing gods.

The Babylonians were responsible for some of the earliest mathematics and they set up the sexagesimal system of angular measurement which is based around the number six and is still used today. They defined a full circle as 360°, corresponding approximately to the 360 days in their calendar; hence 1° corresponds approximately to the angular movement of the Earth during one day in its orbit around the Sun, although the Babylonians did not look at it in that way. They also set up a simple form of algebraic geometry which had many practical uses and formed a basis for surveying.

The Babylonians also initiated the concept of the zodiac, the band in the celestial sphere through which the Sun moves and completes a full revolution in one year. The Moon and planets also move in the same band (but in totally different ways) and we now know that it represents the ecliptic plane in which the planets revolve about the Sun, and the Moon about the Earth. The Babylonians divided it into 12 equal zones of 30° corresponding to the 12 lunar months of their calendar and then assigned each one to the nearest constellation. Constellations were regions of the sky whose stars showed a pattern which could be likened to known people, animals or objects. All the early civilizations identified their own constellations and the 12 signs of the zodiac evolved slowly from ancient roots. The ones that are in use today are those which were recorded by the ancient Greeks; the sequence starts with Aries (the Ram), moves to Taurus (the Bull), then to Gemini (the Twins), to Cancer (the Crab), on to Leo (the Lion), on to Virgo (the Virgin), then to Libra (the Balance or Scales), on to Scorpio (the Scorpion), then to Sagittarius (the Archer), reaching Capricorn (the Goat), then on to Aquarius (the Water Bearer) and to Pisces (the Fishes), which then joins up with Aries to complete the cycle.

During each year, the dates that correspond to each sign of the zodiac are determined entirely by the constellation in which the Sun is located in its migration around the zodiacal plane. The zodiacal year is defined by the time it takes the Sun to complete its full cycle; it is now called the sidereal year, because it is determined by the Sun's movement against the background of fixed stars. In today's terms, it is precisely the time it takes the Earth to complete one full revolution about the Sun. In developing a calendar, the Babylonians knew that a year determined the cycle of seasons, which we now know are caused by the tilt of the Earth's axis of rotation by 23.5° to the plane of its orbit around the Sun. In summer the axis is tilted towards the Sun, causing greater heating because of the longer period of daylight and the greater solar radiation flux per unit area because of the angle of the Earth's surface being less inclined to the direction of the Sun's rays. Mid-summer is defined by the Earth's axis being tilted exactly towards the Sun (the summer solstice), and mid-winter when it is tilted exactly away from the Sun (the winter solstice). (This is the situation for the Northern Hemisphere; the inverse is the case in the Southern Hemisphere.) Spring is defined by the vernal equinox when the Earth's axis is tilted exactly orthogonal to the Earth–Sun line, so that the Earth is equally illuminated from pole to pole and periods of day and night are equal everywhere; similarly, autumn is defined by the autumnal equinox. Hence, a year can also be determined by measuring the direction of the Earth's inclination relative to the Sun, say from vernal equinox to vernal equinox. Such a year, called a solar year, is directly linked to the seasons and is therefore the most important basis for a calendar, the greatest practical value of which is to predict the occurrence of spring, summer, autumn and winter or, more importantly in Egypt, the flooding of the Nile and, in India, the monsoon. Because of this, our modern calendar (whose development I will recount later, starting with the major and leading contribution of the Egyptians) is based on the solar year.

At this point, you may think that the two years I have described – the zodiacal, or sidereal year and the solar year – are identical and simply represent two different ways of measuring the same thing, but this is not so; there is a very small difference between them. The difference is so small that it was unknown to the Babylonians and was not detected until nearly 2000 years after they started studies of the zodiac and their calendar. It is so small that it has a negligible effect on everyday life and, even today, it is not well known outside astronomical circles. However, there is one area of modern activity that should be affected by the small difference between solar and sidereal years but which seems to be largely ignorant of it.

As already said, much of early astronomy was driven by astrology, the belief that human destiny can be foretold in the stars. The zodiac became one of its main tools and, since it is still used today, as can be seen from its presence in the columns of many newspapers, it is appropriate that I bring its story up to date. The simple proposition is to assign to each individual person that zodiacal sign that the Sun was located in when he or she was born. Destinies could then be read at any one time from the positions of the Moon or planets as they moved through the different signs of the zodiac. However, the direction of the Earth's axis is not fixed but, because of its rotation, precesses in a way similar to that of a spinning top; it closely maintains a tilt of 23.5° to its orbital plane but rotates very slowly about it at a rate that will see a complete revolution in 25800 years.

12

Because of this the constellations move slowly westwards along the plane of the ecliptic with respect to the calendar. This is because the calendar is based on the solar year, which, because of the precession of the Earth's axis, is about 20 minutes shorter than the sidereal or zodiacal year. Without precession, these two years would be identical and the zodiacal constellations would not drift through the calendar. Precession was discovered in the second century before Christ by Hipparchus, the most brilliant observational astronomer of ancient Greece. Later, I will describe the method he used, together with his many other achievements.

The Babylonians had adopted the vernal equinox, marking spring, as the start of their year and the zodiac. At the time of the ancient Greeks, the vernal equinox lay in Aries, so this constellation was labelled as the first sign of the zodiac and the vernal equinox is still referred to as the first point of Aries. But when the Babylonians first set up their zodiac, the vernal equinox lay in Taurus; this was therefore the first sign of their zodiac and it accounts for the Babylonian description of Taurus as not only "The Bull of Heaven" but also "The Bull in Front". If our present series of zodiacal signs had been adopted from the earlier Babylonian records rather than the later Greek ones, the first sign of the zodiac would have been Taurus, not Aries, and the astrological discrepancy I am describing would have been even greater. Today the vernal equinox, which occurs on 21 March in our calendar, lies in Pisces; hence, most people who are assigned the astrological sign of Aries should, astronomically, have the sign Pisces. This drift will continue and, during the next millennium, the vernal equinox will enter Aquarius, presumably heralding the "dawn of Aquarius" Hence, if anyone wished to know which constellation the Sun was in on any particular date, past or future, the appropriate astronomical tables would be needed or, more simply, a sidereal rather than a solar year used in which the constellations will not drift. A more simple but approximate calculation can be made from the fact that the constellations drift through our calendar by one zodiacal sign every 2150 years and that the vernal equinox lay in Aries during the great Greek period of a few, say four, centuries before Christ. As of today, this places the vernal equinox in the western part of Pisces.

The astronomical developments in Babylon were led from the temple and were interlinked with religion and the several gods of the time. A primitive cosmology developed, influenced by the nature of Mesopotamia, a land subject to much flooding (as southern Iraq, it is still a marshland today). This said that the gods created the world out of a watery waste and made human beings out of mud to be slaves to them, a world picture appropriate to a society where monarchs had full and absolute power.

Babylon was at its zenith between 1900 and 1600 BC, but for the following thousand years Mesopotamia was like a battlefield, with invasions from all sides. In 1595 BC, the Hittites attacked from the north (modern-day Turkey) and looted and pillaged Babylon; their monopoly of iron gave them military superiority, as well as a great advantage in agricultural tools. In 1100 BC Babylon was conquered by the Assyrians and finally, in 539 BC, it fell to the Persians who established the greatest empire then known through most of the Middle East.

The other great early civilizations, such as those in Egypt, India and China, also conducted astronomical studies which were driven by practical, astrological and religious motives. In China there is evidence that astronomical observations were well under way as long ago as *circa* 2000 BC. The length of the solar year and the lunar month had been determined with good accuracy, allowing a calendar to be developed which could establish the times of festivals and so on. Constellations were identified with emperors, and stars were ranked in status according to their brightness. They had a very simple cosmology: the Universe was the rotating sphere of fixed stars in which the poles were the exalted regions. Astrology was dominant, but little attention was given to the motion of the planets, an important basis of many other astrologies. Instead, changes such as the appearance of a new star or comet were believed to herald calamities on Earth. One of the most dramatic of such celestial changes is a total solar eclipse, and legend has it that two astrologers were executed because they failed to predict one that occurred in 2137 BC. In India, astronomy developed a little later than in China but in a similar fashion. The solar year was measured, a calendar was constructed and the major celestial objects were named after gods and goddesses; astrology was the driving force in these endeavours.

Of course, our knowledge of such ancient activities depends entirely on the evidence handed down to us. The most effective evidence is in written manuscripts, but we must recognize that many of the earliest writings were based on word of mouth transmission through many generations and were therefore written much later (sometimes centuries) after the events they record. Other astronomical developments, of which we do not have a written record, certainly occurred in other parts of the ancient world, and of these we know little or nothing. One example is the megalithic monuments that are found in many countries, especially the British Isles, and which date as far back as the third millennium BC. These represent a considerable engineering feat (some of the stones are as heavy as about 50 tonnes) and their layout, which shows considerable trigonometric knowledge, is clearly dictated by astronomical principles. The most notable of these is Stonehenge in the south of England.

In the above account of astronomical studies in the ancient world, I have given the most detailed description to those in Babylon because they were the most extensive and had the soundest mathematical base. But one other civilization, Egypt, was the first to construct a calendar and, with sub-sequent developments, provided the basis of the calendar we use today. It is therefore appropriate to tell the story of its origin and evolution.

Life in Egypt depended entirely on a great river, the Nile, which flowed through an immense barren desert, devoid of rain. It had no tributaries and it stretched far farther south than the ancient Egyptians ever trav-elled. It flooded every year, bringing a rich soil deposit as well as the water needed for agriculture, making the Nile valley very fertile and allowing the early growth of a great civilization, which was confined to a very narrow strip of oasis, bounded by extensive deserts. Since invaders would have to cross these deserts, they played a protective role and Egypt was not subjected to the same change, invasion and strife that marked the Meso-potamian civilizations of Sumer and Babylon.

We now know, but thanks to the discoveries of the nineteenth century, that the source of the Nile is twofold. One is the great lakes of central Africa which overflow in the aftermath of heavy, regular, annual rains – the White Nile – and the other is the melting of the mountain snows of Ethio-pia, which feed the Blue Nile. The White Nile carries vegetable detritus from the equatorial swamps and the Blue Nile carries ferruginous mud, containing iron compounds, from the soils of Ethiopia. The White Nile starts its 4000-mile course one month before the Blue Nile, and the two streams combine to reach the lower valley just before the summer solstice (mid-June on our present calendar). The inundation floods the whole of the arable land, bringing to the black Earth, burned dry by the Sun, the renewal of its soil and its watering. By late November to early December, the flood has subsided and the wet fertile soil is ready for the ploughshare and seed. Harvesting occurs in March–April, after which the land lies fallow and, without rain, becomes dry and devoid of all vegetation until the next annual flood in mid-June. These events define the three Egyptian seasons of inundation, agriculture and dryness.

Hence, unlike the other early civilizations, the Egyptian "seasons" were not determined by the times of the seasons, but by the Nile and, particu-larly, its flooding, which recurs annually and regularly. It so happens that, when the Egyptians were developing their calendar, the flooding of the Nile was heralded by the heliacal rising of the brightest star in the sky, Sirius (Sothis to the Egyptians), which occurred in mid-June as seen from Egypt. The heliacal rising occurs in the eastern sky just before dawn when the star reappears after having been invisible for about 70 days because of being in daylight. This remarkable event – the reappearance of the brightest star in the sky just before the most important event in Egyptian life, the flooding of the Nile – greatly strengthened the belief in astrology. Other striking astronomical events which recurred (but at that time not predictably), such as the conjunction of two planets, were believed to indicate that the human events of the previous conjunction would be repeated. The Egyptians also intertwined astronomical phenomena with their religion and, like many other ancient civilizations, assigned stars and constellations to different gods and goddesses. Sirius, the herald of flooding of the Nile, was the star of the goddess Isis, consort to the great god Osiris, who was represented by the constellation of Orion. Osiris was originally the god of fertility but was slain by his brother, Seth, who dismembered his body into 14 parts and distributed them throughout the land. This led Isis to embark on a lengthy search, in which she retrieved the parts and reassembled them, bringing Osiris back to life. This familiar ancient story of death and resurrection resulted in Osiris becoming the god of the dead, with the power of salvation or purgatory to each individual in the afterlife. His spirit was believed to dwell in the Pharaoh, thereby endowing him with immense authority as the god incarnate. However, it was believed that, if the Pharaoh died naturally of old age, the spirit of Osiris would also weaken and share the same fate. To avoid this, the spirit had to be released from the body when it was still whole and well, this being accomplished by a religious execution. Not surprisingly, with the passage of time, the Pharaohs adopted the ruse of appointing a substitute who knowingly accepted the spirit and enjoyed the full status and privileges of a king for a few days or weeks before being executed with full ceremonial rites.

Out of these beliefs emerged an explanation of the annual flooding of the Nile on the heliacal rising of Sirius, which was an imaginative and romantic combination of religion, mysticism and astronomy. The heliacal rising of Orion occurs before that of Sirius, so, star by star, Osiris was revealed by this most magnificent of constellations straddling the celestial equator. Then, on the later heliacal rising of Sirius, Isis was also revealed and, seeing her husband, she was immediately saddened and went into mourning for his death. Being a very loving and faithful wife, her sorrow was such as to cause the tears she shed to be so profuse that they inundated the valley of the Nile.

The Egyptians were the first to recognize that the naturally available time intervals of a day, month and year were not commensurable and they were therefore the first to appreciate the difficulty this posed for the construction of a calendar. They took the first major step in developing their own calendar by accurately determining the length of a year as 365.25 days by timing the occurrence of the equinoxes over a long period. They were therefore measuring a solar year, and since this determines the seasons, it is the best basis for a calendar and is the basis of the one we

The magnificent constellation of Orion, followed by the brightest star in the sky, Sirius. To the ancient Egyptians, Orion represented the great god Osiris and Sirius represented his consort, the goddess Isis. In about 3000 BC the reappearance (heliacal rising) of Sirius just before dawn heralded the most important event in the Egyptian calendar – the flooding of the Nile. This was interpreted as being due to the tears of Isis, shed on seeing her beloved Osiris who had been murdered by his brother, Seth.

17

use today. To avoid any confusion, it should be stressed that the small difference between the solar year and the sidereal year, discussed above in connection with the zodiac and caused by the precession of the Earth's axis, has no effect on this story of the development of the calendar. The difference can be detected only by reference to the fixed stars, as was the case with the zodiac, and would ultimately have been revealed to the Egyptians because precession causes the heliacal rising of Sirius to occur later as time progresses. Today, some five millennia later, the rising is no longer in mid-June but in August and it can no longer be regarded as the herald of the flood.

The next important step taken by the Egyptians in constructing their calendar was the decision not to use lunar months, thereby avoiding the cumbersome route to be followed by other early civilizations such as Babylon. They divided the year into 12 months of equal duration, 30 days, and added five intercalary days to make a total of 365 days and thereby harmonize the calendar with the seasons. The start of the year was taken to be 15 June, the official first day of the inundation, and the five intercalary days were devoted to festivals marked by the birthdays of principal deities. Some other days were identified by major historic or religious events and, according to the nature of those events, the days were marked as good or hostile and additional advice given; for example, for one date marking a previous battle the advice was to do nothing, another marking a peace agreement was propitious, and another marking the feast of a god was one in which everything seen would be fortunate. Since we are talking of a time about five millennia into the past, this must represent the first horoscope, as well as the first calendar.

Although the Egyptians knew the length of the solar year accurately as 365.25 days, they did not, at first, introduce an extra intercalary day every fourth year, which would have kept their calendar in step with seasons, the flooding of the Nile and the heliacal rising of Sirius (Sothis). Consequently, they had a sliding calendar through which the above events would move smoothly, completing a full cycle every 1460 years. This is a very long time, but the Egyptian civilization was of very long duration and the period of 1460 years was known as the Sothic cycle because it marked the interval needed for the heliacal rising of Sothis (Sirius) to recur on the first day of the first month of their sliding calendar. The Egyptians were therefore fully aware of the true solar year from the timing of the equinoxes, solstices and the reappearance of Sirius, so they effectively had a seasonal calendar (also called the Sothic calendar) allowing the prediction of the above events within their sliding calendar. Carrying two calendars had an advantage, in that the difference between the two told what year it was since the time when the two were previously synchronized (one day difference represented four years). We know from historic evidence that they were in step in AD 139 when the heliacal rising of Sirius occurred on the first day of the sliding calendar. Since this synchronization happens every Sothic cycle of 1460 years, it also occurred in 1321 BC, 2781 BC and 4241 BC; hence, one of these dates probably marks the date on which the Egyptians first set up their calendar. An even earlier date can be ruled out from archaeological and other evidence. Some historians favour the earliest of these dates and, if so, 4241 BC represents the first accurately dated year in history, but 2781 BC seems more likely.

The Egyptians broke the day into 24 parts, thereby creating the basis for today's 24 hours. They assigned 12 parts for daylight and 12 parts for night, and measured the former from the direction and length of the Sun's shadow and the latter from the positions of stellar constellations which they defined for the purpose. Since the duration of day and night varies through the year, this meant that an hour also varied from day to night and through the year. A constant hour was not established until the development of accurate mechanical chronometers some thousands of years later, the first of which were based on the pendulum. But credit for the very first known man-made clock lies with the Egyptians, who supplemented their solar and stellar time measurements, which could only be conducted when the sky was clear, with a water clock which could be used at any time. This consisted of a large vessel from which water leaked through a small aperture at a constant rate. The vessel was translucent, enabling the level of the water to be seen and measured against a graded scale on the outside.

The Egyptians did not abandon their sliding calendar in favour of the much better Sothic calendar, but maintained it as the official calendar, probably due to the opposition of priests to any change. In the aftermath of the conquests of Alexander the Great, there was a strong Greek presence in Egypt, mainly centred on the city of Alexandria, founded in 322 BC. Acting on advice by Aristarchus, a brilliant Greek philosopher and astronomer (of whom more later), the then King of Alexandria (Ptolemy III) proposed in 238 BC the adoption of a seasonal calendar of 12 months of 30 days, plus 5 intercalary days, plus an additional day every fourth year, like our own present-day calendar. However, opposition to this change was strong in Egypt and it was not implemented, the sliding calendar being maintained for establishing the dates of festivals and religious events, and the Sothic calendar for determining the seasons.

In the early days of its empire, Rome adopted a calendar essentially based on that of Babylon, nearly two millennia before. There were 12 months determined by the first appearance of the new Moon, and therefore of 29 or 30 days' duration, giving a year of 354 days. Since this rapidly got out of step with the seasons, it was corrected, as in Babylon, by the insertion of an additional month every three years or so. Like the

Babylonians, the Romans started their year at the vernal equinox and named their months by number. Some of these still survive in our calendar today, where September, October, November and December are derived from the Latin words for seven, eight, nine and ten. Of course they do not correspond to those numbers in our calendar today because the first month of the old Roman calendar coincided approximately with our March.

The Roman calendar, based on lunar months, was very unsatisfactory and untidy, and could give different dates in different parts of the empire, because of the uncertainty in identifying the day of the new Moon. Accordingly, Julius Caesar decided to introduce a new calendar and, having been to his wars in Egypt and having noted the advantages of its calendar, he invited Sosigenes of Alexandria to advise on its construction. Sosigenes proposed the same calendar as that proposed by Ptolemy III nearly a hundred years before – 12 months of 30 days, plus 5 intercalary days, plus an additional day every fourth year. He also proposed that the year should start with the five intercalary days or, in a leap year, six. These proposals were modified slightly in Rome; the five intercalary days were spread through the year with months being formed alternately of 30 and 31 days. Since this would have made 366 days, one day too many in a year, one month (February) was assigned one day less (29) but would have 30 in a leap year. (Retrospectively, it would have been tidier to remove the day from a month with 31 days.) Sosigenes had also proposed to start the year at the winter solstice and after the intercalary days the first day of the first month would begin. This was accepted but, in an oversight, no account was taken of the dispersal of the intercalary days throughout the year, with the result that the first day of the first month, which now became the first day of the new year, started five days after the winter solstice and placed the vernal equinox at 25 March. This Julian calendar was introduced in 46 BC, two years before Caesar's assassination, but a small modification was made by Augustus who became the first Roman emperor in 27 BC. The seventh month had been named July in honour of Julius, so it was natural that the next month be named August after Augustus. But July had 31 days and August only 30 and, since this may have been taken as a measure of their relative greatness, Augustus stole another day from February (giving it 28) and added it to August to make it 31 and therefore equal to July. In order to resume the 30–31 sequence, the subsequent month, September, was assigned 30, and so on. This explains the somewhat illogical distribution of days through the months of our present calendar.

The Julian calendar was used throughout Europe until AD 1582, when an adjustment was made by Pope Gregory XIII to create the Gregorian calendar. The solar year is not exactly 365.25 days, but is about 11 minutes shorter. This is not much, but in the 16 centuries since the initiation of the

Julian calendar, it had built up to 14 days and it was resolved to correct for this and bring the calendar back into step with the seasons. To do this would require the removal of 14 days, but the Church, for reasons connected to the timing of the great Christian festival of Easter, which was dated by the Jewish lunar month calendar, wished the vernal equinox to be dated as 21 March and not the 25 March of the Julian calendar. Hence, the required adjustment became ten days and Pope Gregory decreed that, in the year AD 1582, ten days would be removed from the month of October and 4 October was followed by 15 October. This adjustment of the vernal equinox by four days also moved the winter solstice backwards in the calendar by the same amount. These four days, added to the five days caused by the error in not accounting for the removal of the intercalary days from the start of the Julian calendar, explains why, in our present calendar, the first of January is nine days after the winter solstice.

In order to avoid any future adjustments to the calendar caused by the solar year being slightly less that 365.25 days, a fine tuning of the leap-year rule was introduced in which century years not divisible by 400 would not have an extra day added, even though they meet the normal four-year rule. Hence, 1700, 1800 and 1900 were not leap years but 2000 will be. The Gregorian calendar was immediately adopted by all Roman Catholic European countries, but in those early post-reformation years, the Protestant states of northern Europe and the orthodox Christian countries of eastern Europe did not follow suit. Britain waited until 1752, when 11 days had to be deleted from the Julian calendar, and Russia did not change until the twentieth century, when 13 days had to be deleted. This explains why the Bolshevik revolution of 1917 is called the "October" revolution; it fell in that month on the 25th in the Julian calendar although it now falls on 7 November in the Gregorian calendar.

Although two developments in ancient astronomy, the zodiac (see Colour Plate 1) and the calendar, have been brought up-to-date, this narrative is only approaching the first millennium before Christ. During that time, as has been stressed, the driving force for studying the sky came from astrology, with astronomy playing a very subsidiary role but benefiting from the fall-out. From then on, this situation was to change, albeit slowly, with astronomy playing a greater and greater role relative to astrology and, although the coupling of these two completely incompatible activities weakened with time, it was not finally broken until the closing stages of the Renaissance. But, as far as this story goes, the connection is being severed now because it is relating the human attempt to understand the Universe we live in, something to which astrology, apart from the ancient stimulation of astronomical observations, has made no contribution. No further reference will be made to it, and the author defers to Shakespeare:

*This excellent foppery of the world, that, when we are sick in fortune –
often the surfeit of our own behaviour – we make guilty of our disasters
the Sun, Moon and the stars, as if we were villains by necessity, fools
by heavenly compulsion, knaves, thieves, and treachers by spherical
predominance, drunkards, liars, and adulterers by an enforced obedi-
ence of planetary influence; and all that we are evil in, by a divine thrust-
ing on.* **King Lear**

CHAPTER THREE

The Greeks

From harmony from heavenly harmony
This universal frame began: . . .
Through all the compass of the notes it ran,
The diapason closing full in Man.
Dryden

If, intellectually, the fourth, third and second millennia before Christ belonged, in different measures, to the great civilizations of Mesopotamia, Egypt, India and China, the first undoubtedly belonged to a small country in southeast Europe: Greece. It was not a single country, but consisted of various groupings and city states such as Macedonia, Athens and Sparta, which were often at war with each other. But the bond that held them together as a whole was the Greek language, which extended all around the Aegean. In this region, starting about seven centuries before Christ and to last for several centuries, the most remarkable human intellectual explosion occurred, the like of which had not been seen in any previous civilizations, and the like of which was not to be seen again for 2000 years. There were many major and several seminal developments in every area of scholarship – in literature, history, drama, architecture, art, sculpture, politics, philosophy, mathematics and science. The great ancient Greeks are household names today, as are the characters in the first great works of literature and history – the epics of the Iliad and the Odyssey, written by Homer about the siege of Troy and the wanderings of Ulysses, events that occurred some centuries previously, *circa* 1200 BC, accounts of which had been passed down orally in the form of ballads, presumably with some embellishment. The great achievements of ancient Greece were to dominate human thought in the West for two millennia and they still have an influence on many present-day activities. Even in athletics, we have a major legacy, with one of the greatest sporting occasions in the calendar, the Olympic Games, deriving from the first games held by the Greeks in Olympus in 776 BC.

The dominance of ancient Greece was intellectual and cultural and, with one notable but brief exception, was not military. In the middle of the first millennium before Christ, the Persian Empire, the greatest the world had known, stretched throughout the Middle East and bordered Greece, causing enmity and resulting in many military conflicts, for example, the battle of Marathon in 490 BC. In 338 BC, Phillip II of Macedonia defeated other states in Greece to become dominant over the whole region. His assassination two years later led to the succession of his son Alexander, who raised an army and crossed into Asia at the age of 22. He regarded his expedition as a Hellenic crusade against Persia, and the combination of a brilliant military mind and near-reckless courage led to a remarkable series of victorious battles, often against superior numbers, by which he conquered Persia, Egypt, and even reached the Indus valley in India, where his army, bored with victories, refused to go farther. He returned to Babylon, where he died at the age of 32, ten years after leaving Greece. The great heritage of Alexander the Great, who was tutored by Aristotle, was to be intellectual and cultural, since his empire soon divided and then collapsed, but the Greek language and culture survived. Alexander had set up city states on the Grecian model, and one of these, Alexandria, founded in 332 BC on the Mediterranean coast of Egypt, was to become a centre of Greek learning

But my story concerns the pursuit of astronomy and I have to select from the great Greek achievements those relevant to that subject. I start with the work of Thales (*circa* 629–555 BC) who is credited as being the first philosopher. At that time, philosophy included science, for which there was not a separate word, and being a philosopher usually meant being a scientist; in the case of Thales, we would call him an astronomer. He was born in Miletus on the Greek coastal fringe of Asia Minor and founded the Early Ionian School, which marked the start of Greek science.

In order to learn the techniques of astronomical observation, Thales travelled to the great centres in Babylon and Egypt where he also learned the art of surveying and used this to measure the heights of the pyramids. His return marked the introduction of astronomy and elementary trigonometry into Greece. He is called the first philosopher because he introduced the first spirit of true enquiry into the nature of things. "What is the Universe?", "What is life?" are questions which had been asked for centuries before but had been answered with myths that were usually irrational, subjective, specific and dogmatic. Greek philosophy tried to give rational, impersonal answers that could be subjected to criticism and query.

Thales gave the first general description of nature and the Universe without invoking the supernatural. It was a very primitive and wholly inadequate concept, but in its spirit it was very new. He had noticed that there were three states of water: not only the liquid state, but also solid, in the form of ice, and gaseous in the form of water vapour. He proposed that everything, including the Sun and stars, were made by some different form of water. He noted that living things perish without water and argued that this was because, being a form of water, they needed replenishment. He proposed that the Universe was a gigantic ball of water in which our world existed in a bubble within which the Earth floated on a water base at the bottom and there was also water above from which the rains came. The heavenly bodies float in the universal water and move in a manner to explain the observations. Previous astronomers (astrologers would be a more accurate term) regarded the stars as Gods; to Thales they were made of a familiar material – water in its different forms. Ludicrous though this proposition might seem, it has to be put into the context of the time and it represents the first non-occult attempt to describe the Universe and its composition, an exercise that still continues today. Thales could also be called the first absent-minded professor. He is reported by Plato to have been so absorbed in his astronomical observations that on one occasion he fell down a well – an incident that was to be commented on centuries later in a quite different context (see page 44).

The next person in the story is Pythagoras (*circa* 580–500 BC) who was born in Samos, an island just off the coast of Asia Minor and not far from Thales' birthplace of Miletus, on the coast. The local ruler was something of a tyrant and this led Pythagoras to leave Samos and, after some travelling, he settled in Croton, a Greek colonial city-state, in the very south of Italy, where he set up a school that became famous and long-lived. The great Grecian intellectual developments occurred in different parts of the Greek-speaking regions, and later major schools were to develop in Athens and Alexandria. The Pythagorean school had a mystical and religious element, and included a belief in transmigration and reincarnation. However, under Pythagoras, there was an astonishing development in arithmetic and geometry, leading to the just claim that he was the first real mathematician. He linked numbers (arithmetic) to shapes (geometry) and ultimately to nature and to astronomy. His linking of numbers with nature was first made via musical harmony. He pointed out that the note from a plucked taut string was only harmonious when an exact number of (standing) waves were excited. Hence, when only the end points are stationary (nodes), there is half a full wave and the chord emits the ground note. When it is a full wave, or two half-waves, the frequency has doubled and the note is an octave higher. When it is three half-waves, the note is one-fifth above the first octave, when it is four half-waves, it is an octave above the first octave, and so on.

Pythagoras then made the proposition that nature was harmonious and could be represented by numbers, that is, by whole numbers or their fractions: 1, 2, 3, $\frac{1}{2}$, $\frac{3}{4}$, and so on, and he applied this principle to his view of the Universe. He believed, as did other philosophers, that the heavenly bodies were carried around on spheres. This belief arose naturally from watching the night sky, where the stars looked like a sphere rotating about its apex at the north pole. Pythagoras proposed that the spheres carrying the Sun, Moon, planets and stars around the Earth would be harmonious and could be represented by numbers. Subsequent attempts at variations on this theme have been made, but all without success. But the great thought that Pythagoras had was to try to link nature and mathematics. It was a thought that was not taken up by other Greek philosophers; indeed, mathematics and philosophy were to develop independently. The 13 books of Euclid's epic *Elements*, which formed the basic textbooks on geometry for more than 2000 years, were entirely mathematical treatises, and the astronomical developments I will relate shortly used mathematics only as a supporting tool and not as a means of explaining nature. This separation of mathematics and natural philosophy was to continue until the monumental work of Isaac Newton.

The two great schools of Thales and Pythagoras were now to be joined by another, possibly the greatest, in Athens. This was founded by Plato (*circa* 427–347 BC), an Athenian and disciple of Socrates who, after travelling widely, formed his school in 368 BC. He called it the Academy, after the name of an olive grove where he used to teach and which derived its name from an earlier Greek hero, Academus. The name "academy", from which we also derive the term academic, is still with us today and is a name given to institutes and organizations of great learning, not only in science but also in the arts, music and literature. Plato's contributions to philosophy (as we understand the term today) and to theories of ideas, knowledge, morality and politics, were immense. In natural philosophy and astronomy, he believed that the material world we lived in was an imperfect representation of the true Universe and, therefore, that we could learn more about that Universe by logic and reason than by observations, since these could only give an incomplete or defective picture. On this basis he argued that, since the sphere is the perfect geometric shape, all astronomical bodies must be spherical and their motions circular. But his greatest contribution to science was the creation of the Academy, where one of his most gifted pupils, Aristotle (*circa* 384–322 BC), was to follow in his footsteps and lay down principles of the nature of the Universe that were to go unchallenged for 2000 years.

Aristotle was born in Macedonia on the northwest coast of the Aegean. He went to Plato's Academy at the age of 18, staying there for 20 years until Plato's death, when he left and returned to Macedonia for a few years in order to educate King Phillip's son, Alexander. There is no doubt that, as the leading Greek philosopher, his revelation of the latest intellectual developments to the young prince contributed markedly to Alexander's greatest legacy: the spread of Greek culture through the Middle East and into Egypt, where the city, Alexandria, founded in his name, was to rival Athens as a centre of learning. Aristotle returned to Athens in *circa* 335 BC, where he opened his own school, the Lyceum. But on the death of Alexander (now called the Great) in 323 BC, anti-Macedonian feeling forced him to leave Athens and he went to the Greek island of Euboea, where he died.

Aristotle had a remarkable intellectual energy and a curiosity of the highest order. His interests spread over all branches of learning and, being a prolific writer, his contributions to many fields were very great; these include psychology, morality, politics, biology and, finally, cosmology, the area I will address shortly. Aristotle's influence was immense, possibly greater than any other philosopher and for a very long period of time. In the Middle Ages, he was referred to as *The Philosopher* and his works were translated extensively and were accepted without question, being enshrined in the belief of the certainty of ancient writings that developed at that time. This attitude is neatly represented by Pope's eighteenth-century poem, the "Danger of Imperfect Knowledge", from which the following extract is taken:

Concluding all were desprate sots and fools,
Who durst depart from Aristotle's rules.

Aristotle's writings on natural philosophy and astronomy were not entirely his own, since he drew on the work of previous philosophers, not only Greek. But he was the one who documented them and he added his own distinctive ideas, which were based on *a priori* reasoning rather than experiment or observation. This approach was quite contrary to his studies in biology, which were dominated by observations; in his four principal writings on animals, he gave detailed descriptions of their structure and movement, and also placed them into classifications, the first attempt to do so. But in astronomy, his belief, derived from Plato, was that the power of the intellect was sufficient to determine the nature of the Universe and, if the logic was impeccable, what was the need for experimental test? There is no denying the high degree of Aristotle's intellect, but even it was insufficient to justify his ultra-confidence in logic over experiment; as today's science tells us, the two go hand in hand, with the former following the latter.

Aristotle's cosmology was very neat and tidy. The Universe was finite and bounded by a sphere that contained the fixed stars (see Colour Plate 2). Outside the sphere there was nothing – no body, no place, no time.

Between this sphere and the Moon's orbit was the intermediate region containing the Sun and planets, with the Earth and the Moon forming the inner region. The Earth was stationary and at the centre of the Universe. The three regions were different in their nature, as well as in their location. The inner Earth–Moon region was subject to change, as clearly evidenced on Earth by thunderstorms, precipitation, rainbows, flood, and so on, and therefore all phenomena that change are in the inner region; this clearly identified comets as sublunar. In the inner region, Aristotle proposed that everything was made from four elements – earth, water, air and fire – but this first attempt at chemistry was not original, having been developed earlier in India and having a certain sense to it in that the first three represented the three states of matter that we are familiar with on Earth – solid, liquid and gas – to which fire had been added. The stars, at the boundary of the Universe, were quite different; they were fixed, unchanging, divine and eternal, a view held by Plato, which clearly influenced Aristotle. The stars therefore had to be made of something quite different to anything existing on Earth and he proposed that they were made by a sublime element, ether, which had the property of self-luminescence, thereby explaining the light emission from them. His creation of this *fifth* element is the origin from which the word *quintessence* derives. The ether is also present in the intermediate region below the boundary containing the stars, but the Sun and planets suffer contamination as the Moon is approached.

Aristotle also recorded the first known views on motion and gravity. He stated that the natural state of a body was stationary: nothing moved unless it was pushed. On Earth the four elements had an inherent force that drove them towards their natural place in the world and, on arrival, the cause of motion disappears and becomes the cause of quiescence. A solid object (made of the first element, earth), on being dropped, will fall vertically in a straight line to its home, the ground, where it becomes stationary; a heavy object will fall faster than a light object, twice as fast if it is twice as heavy. Of course, the object can be picked up and hurled back into the air, but its inherent force will eventually overcome that motion and it will return to the ground and again become stationary. Aristotle argued that if a galloping horseman threw a stone straight above his head, it would return vertically downwards towards the ground and, if the horse was travelling fast enough, the stone would fall behind the rider. Unfortunately, he did not conduct this experiment, which would have proved him wrong.

The inherent source of motion in the second element, water, also caused heaviness, so it would also fall towards the ground, but, in the case of rain, a wind would push the waterdrops from their chosen vertical paths, causing them to fall obliquely. On reaching the ground, however, the water had still not reached its destination and it would flow via streams and rivers into the seas and oceans where, having reached home, its internal force is satisfied and it becomes stationary except for motions such as waves caused by the pushing of the winds. The third element, air, has the property of lightness and it is at home where it is, immediately above the land and seas. Of course, it moves in the form of winds but these must be caused by some unknown force. The lightest element of all is the fourth, fire, whose inherent force drives it upwards to its natural home above the atmosphere; this could be seen in the upward flames from any fire.

In considering the motion of the heavenly bodies, Aristotle reasoned that the laws governing them must be different from those on Earth. On Earth, the natural state is to be stationary, with all motions ultimately ending, but the motions of the heavenly bodies were permanent and endless. What could be the nature of permanent motion? Aristotle argued that it could only be motion that was unchanging in every respect and he concluded that this meant circular motion at constant speed. Unlike a straight line, there was no beginning and no end, and there was no change in velocity. It also fitted in neatly with his notion (and the notion of many previous philosophers, including Pythagoras) that all the heavenly bodies, the Sun, Moon, planets and stars, were carried on spheres which pushed them around in their orbits. The motion, therefore, would be circular and the spheres would need to rotate at constant speed.

These views of Aristotle on natural philosophy and cosmology were to last for 2000 years, a record for any scientific concept or hypothesis. Subsequent civilizations were to translate his writings meticulously into their own language and accept them without question until that second great period in human intellectual activity, the Renaissance, when they were demolished – but not without pain, difficulty and controversy.

So far, Greek astronomy had been largely qualitative, but it now entered a quantitative phase, represented largely by the work of Aristarchus, Eratosthenes, Hipparchus and, finally, Ptolemy.

Aristarchus (*circa* 310–230 BC) of Samos was one of the last of the Pythagorean School. He became a great scholar in Alexandria, where he advised Ptolemy III on the Greek calendar as described earlier. Alexandria had become a great centre of Greek learning, rivalling Athens. It was ruled by the Ptolemaic dynasty which had started with Ptolemy I, one of Alexander's best generals, and continued until the Roman conquest in 30 BC. Aristarchus was not the first to ask the questions – How far is the Moon? How far is the Sun? How far are the stars? – but he was the first to

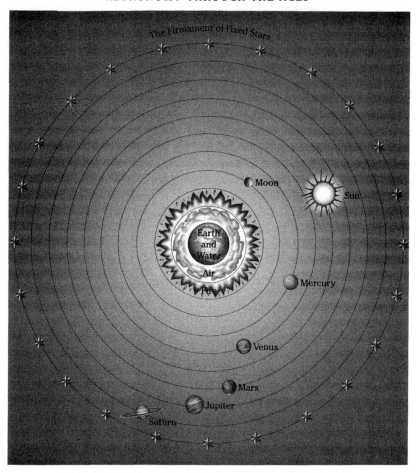

Aristotle's view of the Universe with the Earth at the centre, composed of earth and water, surrounded by air and fire and then, in order of distance, the Moon, Mercury, Venus, the Sun, Mars, Jupiter, and, finally, the firmament of fixed stars composed of the fifth element, ether.

devise geometrical methods to answer them. So, he was the first to attempt the determination of the distance of heavenly bodies, one of the most important and basic tasks in astronomy and one that is still a major challenge today; for example, we are still not sure of the size of the Universe to much better than a factor of two.

The methods of Aristarchus were geometrically correct, but the precision of measurement then available was such as to greatly limit the accuracy of his estimates. Nevertheless, he was able to show that the Sun was much farther away from the Earth than the Moon was and that it was much larger than the Earth, which was itself larger than the Moon. He was also able to give an indication of the immense size of the then Universe – the distance to the visible stars.

For the Earth–Moon system, Aristarchus used the eclipse of the Moon by the Earth. In such a lunar eclipse the Sun, Earth and Moon are in line in that order, with the Moon appearing full. But when the Moon passes into the shadow of the Earth, the relative sizes of the two bodies can be estimated from the curvature of the Moon's bright disk and the curvature of the Earth's dark shadow on it. Aristarchus concluded that the Earth was three times larger than the Moon; it is actually nearly four times larger, but the error of about 20 per cent is a reasonable one for that time and that method. Since the angle that the Moon subtends to the Earth can be measured (it is about half a degree), its distance can then be estimated in Earth diameters; Aristarchus obtained 25 Earth diameters, compared to the modern value which is close to 30.

Aristarchus then turned to the much more difficult problem of the size and distance of the Sun. He pointed out that, when the Moon was exactly half illuminated by the Sun, the angle between the Earth–Moon line and the Moon–Sun line was a right angle (90°), so if he measured the angle between the Earth–Moon line and the Earth–Sun line, he could calculate the distance of the Sun in terms of the Earth–Moon distance, which he knew in units of Earth diameters. The method is geometrically correct, but the precision required for its use is extremely high because of the very great distance of the Sun compared to the Moon, which makes the angle to be measured very close to a right angle itself; further, the determination of when the Moon is exactly half full is not easy. Aristarchus made his measurements in the best and most accurate way he could. He measured the times at which the Moon was at first quarter (half full and waxing) and at third quarter (half full and waning). The time interval from the third to first quarter, when divided by a whole lunar month, represents the fraction of a full circle made by twice this angle. He obtained 87°, which meant that the Sun was 20 times more distant than the Moon. Also, since total solar eclipses showed that the angular diameters of the Sun and Moon were the same, this meant that the Sun was 20 times larger than the Moon or, with Aristarchus' estimates, seven times larger than the Earth. These values are grossly inaccurate; the actual angle is very close to 90° (89.9°) and the Sun is nearly 400 times more distant than the Moon, and more than a hundred times larger than the Earth (more than a million times larger in volume). But Aristarchus had demonstrated that the Sun was more distant than the Moon and larger than the Earth, a considerable achievement for the time.

These methods were relative to the then-unknown diameter of the Earth, but in *circa* 250 BC a young mathematician arrived in Alexandria whose work was to allow Aristarchus to calibrate his measurements. The young man's name was Eratosthenes (*circa* 273–195 BC), who was born in

Athens but was to spend most of his life in Alexandria, whence Ptolemy III summoned him in order to tutor the king's son. Eratosthenes' great and remarkable achievement was to measure the size of the Earth.

Eratosthenes had noticed that at Syene (present-day Aswan) the Sun was directly overhead at noon in mid-summer. Being a mathematician and skilled at geometry, he devised an experiment to measure the size of the Earth, assuming, as proposed by Pythagoras, that it was round. He erected a tall vertical pole (a gnomon) at Syene and a similar one in Alexandria. At the summer solstice, the Syene gnomon cast no shadow, showing that the Sun was directly overhead, but that in Alexandria cast a small shadow whose length was one-fiftieth of the height of the pole. The angle involved (about 7°) represents the angle between the two lines drawn from the Earth's centre to Syene and Alexandria respectively, and simple geometry says that the distance between them is 2 per cent the circumference of the Earth. The unit of length used at the time was the stadium, and Eratosthenes measured the Syene–Alexandria distance to be 5000 stadia, giving the Earth's circumference as 250000 stadia. It is believed that one stadium equalled 160 metres, giving the Earth's circumference as 40000 kilometres (about 25000 miles). Since Alexandria is due north (closely) of Syene, Eratosthenes was measuring the *polar* circumference of the Earth (of course, he did not know that the Earth was not exactly spherical) for which the present value is 39940 kilometres. The measurement of Eratosthenes is therefore quite remarkable in its accuracy.

The only writing of Aristarchus that has survived is the one describing the above methods for determining the sizes and distances of the Moon (see Colour Plate 3) and Sun. But he was very active in many other areas of astronomy as references by other writers indicate. In particular, Archimedes, who established the law of displacement and the principle of the lever, briefly refers to Aristarchus' cosmology, which was magnificent in its intellectual daring. Aristarchus was aware of Aristotle's cosmology, but proposed a quite different system which shed most of Aristotle's complexity. He had shown that the Sun was much larger than the Earth (more than 300 times the volume, on his own estimate) and therefore considered it a much better candidate than the Earth for the central place in the Universe. He accepted Aristotle's bounded Universe of a sphere containing the fixed stars, but made it stationary and placed a stationary Sun at its centre. The Earth was no longer the centre of the Universe, but revolved about the Sun in a circle at constant velocity and rotated about its axis. Following his passion for determining astronomical distances, he realized that his model gave him a method of measuring the distance to the stars because the motion of the Earth from one point in its orbit to the extreme opposite point would cause the stars to show parallax, that is, they would appear to be slightly shifted in the sky. But he was faced with

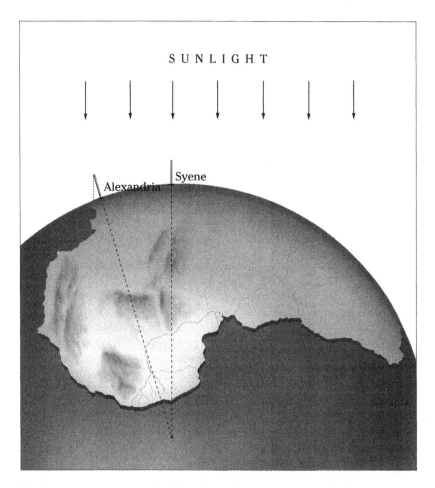

This demonstrates the method that Eratosthenes used in the third century BC to measure the size of the Earth. In midsummer a vertical pole (a gnomon) at Syene (modern day Aswan) cast no shadow at noon, whereas a similar vertical pole in Alexandria cast a shadow, which was one 50th of its height. Simple geometry meant that the distance from Syene to Alexandria was one 50th of the Earth's circumference. This gave the circumference to be 40000 kilometres, a remarkably accurate result.

an even greater problem than in his attempt to measure the distance to the Sun, because the stars are immensely distant, even compared to the Earth–Sun distance. No parallax could be detected, leading him to conclude that the size of the Universe (at that time the distance to the fixed, bright stars) was so great that, as he said, if you drew a very large circle, the Earth's orbit would only appear as a point. The non-detection of stellar parallax, as Aristarchus realized, was attributable to the great distance of the stars. He did not know it then, but even the very nearest stars only have a parallax of about one second of arc and it was not until the nineteenth century that this was finally detected.

The cosmology of Aristarchus was not accepted in the ancient Greek world; Aristotle remained supreme. One reaction we know about is that the Stoic, Cleanthes, proposed that Aristarchus should be indicted for impiety for his heliocentric views. Whether this was known eighteen centuries later to an obscure Polish canon of the Christian church who was to propound the same concept (but in a quantified form) or to an ebullient Italian scientist who was to defend it against a formal charge of heresy, is not known. Their names were Nicolaus Copernicus and Galileo Galilei; I shall consider their contributions later.

Much of the astronomical analysis during the Greek period that has been dealt with so far was based on the extensive, but rather crude, earlier observations of Babylon. This situation was to be changed dramatically by one man, Hipparchus (*circa* 194–120 BC), who can be said to be the first really great observational astronomer. He was born in Nicaea (now Iznik on the coast of Turkey, an area that produced many of the great Greek philosophers) but spent most of his life on the island of Rhodes and in the city of Alexandria. He conducted astronomical observations from Rhodes over a 30-year period, using a consistent method based on a simple instrument, a rod, which could be swivelled about the vertical and was pointed at the object being observed; the angle to the vertical was measured against a circle marked out in degrees. He is also credited with the invention of the astrolabe, a portable instrument to measure the altitudes of celestial bodies, based on the same principle and which is the forerunner of the sextant. Over a period of about 30 years, Hipparchus mapped the positions of 1080 stars and gave estimates of their brightness; he did the same for the five planets: Mercury, Venus, Mars, Jupiter and Saturn. His data were to be the basis for astronomical studies for the next 17 centuries, before they were finally superseded by those of the second great observational astronomer, a Danish nobleman called Tycho Brahe.

As was related earlier, in discussing the zodiac, one of the great achievements of Hipparchus was the detection of the precession of the Earth's axis, which maintains the same angle to the plane of its orbit but rotates about the line at right-angle to that orbit with a period of about 26000 years. So, it is very slow, but it has the effect that the solar year, which determines the seasons and only depends on the direction of the inclination of the Earth's axis to the Sun, is slightly shorter (about 20 min-

34

utes) than the sidereal year, the time for the Earth to complete a full orbit about the Sun, which is determined by the stars. He did this by brilliantly exploiting the advantage of a long time base in astronomy. Using measurements of the equinoxes 150 years previous to his own, he was able to detect the much greater precession, 150 times more than in one year, of 2°. Effectively, he had determined accurately the duration of the solar and sidereal years, but he did not understand it in those terms.

Hipparchus also sought to bring to geography the precision of astronomy by using his stellar observations to locate places on Earth. Hence, he transferred his astronomical coordinate system of right ascension and declination to their geographic counterparts of longitude and latitude, which we still use today. He also applied the method of Aristarchus to determine the distance to the Moon and arrived at a remarkably accurate value of 30 Earth diameters. Using the equally remarkable measurement of the Earth's diameter by Eratosthenes, he derived the Earth–Moon distance to be 382000 kilometres, accurate to about 1 per cent! Although Hipparchus was aware of the heliocentric view of Aristarchus, he favoured the geocentric view of Aristotle because of the non-detection of parallax (caused, as I have said above, by the immense distance of the stars). Using his own data, he attempted to quantify Aristotle's model, but soon realized, as others had before, that a simple, single sphere could not explain the motions of heavenly bodies, particularly those of the planets. He introduced the concept of secondary circular motion on a sphere rotating around a larger sphere. These "epicycles" were to be adopted, some 300 years later, by the last of the great ancient Greek philosophers, Claudius Ptolemy of Alexandria, in his epic quantification of Aristotle's cosmology using the observations of Hipparchus.

Not much is known about the personal life of Claudius Ptolemy of Alexandria (*circa* AD 100–170), but his scientific works are well known because his writings survived. His *magnum opus*, the main result of his life's work, is known by its Arabic title (a language it was translated into in the ninth century): the *Almagest* or "the Greatest". Looking back, this is not quite as conceited as it seems. It was an epic that was to be used for 14 centuries to predict the positions of the Sun, Moon, planets and stars to a precision of about 1°, the accuracy of the observations of Hipparchus.

The *Almagest* consisted of 13 books. The first presented the geocentric cosmology of Aristotle with arguments in its favour; the second gave an outline of the spherical trigonometry he was to use to quantify that cosmology; the third discussed the motion of the Sun and the length of the solar year; the fourth discussed the Moon's motion and the length of the lunar month; the fifth dealt with the distances of the Sun and Moon; the sixth outlined the principles of eclipses of the Sun and Moon; and the seventh and eighth books dealt with the fixed stars, giving a catalogue

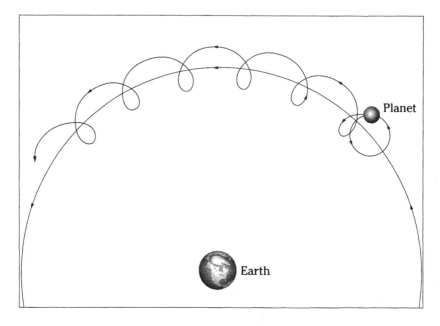

Ptolemy's world model was based on the cosmology of Aristotle, which constrained the motion of all heavenly bodies to be circular at constant velocity. In order to explain the motion of the planets, Ptolemy had to introduce secondary circular motions (epicycles) whose centres moved in circles about the Earth.

which was largely based on Hipparchus and including a discussion of the phenomenon of precession. These eight books were essentially a description of the achievements of Greek astronomy before him, but the last five books represented his own major contribution to astronomy. He took the cosmology of Aristotle and quantified it by using the observations of Hipparchus, supplemented by some of his own. This was an immense task, which he completed successfully, but not without difficulty.

The geocentric cosmology placed the stationary Earth at the centre of the sphere, which carried the fixed stars and marked the boundary of the Universe. All the other celestial bodies, the Sun, Moon and planets, were also carried by spheres which caused them to move in circles at constant velocity. The observations showed that the stars moved from east to west during the night and that new constellations would appear just before dawn (heliacal rising) and rise earlier and earlier each night until they passed over the whole sky, disappeared into daylight and reappeared in a heliacal rising one year later. The Sun, Moon and planets all had their individual periods of change, but, against the background of fixed stars, the motion of all of them was confined to the ecliptic plane – the band that carried the constellations of the zodiac. The Sun rose and set every day with a period of 24 hours, and its inclination at noon varied from its highest in mid-summer to its lowest in mid-winter (Northern Hemisphere) causing the seasons to change over one year. The inner planets (Mercury and Venus) were seen only when near the Sun, just before sunrise or just after sunset. The outer planets (Mars, Jupiter and Saturn) tended to move eastwards against the background stars, but could also reverse this motion and move backwards; these retrograde motions, indeed the planets as a whole, were to present Ptolemy with his greatest challenge.

Earlier astronomers, such as Hipparchus, had recognized that planetary motions could not be represented, even crudely, by a simple circular motion about the Earth, and had proposed a combination of two circular motions, one in a large circle (deferent) and one in a smaller circle (epicycle): wheels within wheels. The centre of the epicycle moved around the deferent at constant velocity and the planet moved around the epicycle at a constant but different velocity. Ptolemy adopted this approach and, by adjusting the sizes and velocities of deferent and epicycle, he was able to explain retrograde motion. For the inner planets, he also needed epicycles, but to explain the fact that they were always close to the Sun, he had to make the centres of their epicycles lie on their deferent at the point intersected by the Earth–Sun line. Of course we now know that the retrograde motion is an apparent one caused by the faster Earth catching up with the outer planets and overtaking them, and that the inner planets are never seen far from the Sun because they actually are near to the Sun.

At this point, Ptolemy had a good model of the geocentric Universe but not, in his eyes, good enough, because it did not match the accuracy given by the observations of Hipparchus. He therefore adopted what today we would call a fudge, and he introduced two slight deviations from Aristotle. First, he allowed the Earth to be slightly displaced from the centre of the planetary orbits and, secondly, he allowed a small variation in the velocity of the planets, as seen from Earth, by establishing another point (the equant), which was also displaced from the centre by the same amount as that of the Earth, but on the other side, from which the angular velocity of the planet was constant. These adjustments, which were quite arbitrary and had been introduced only to match the observations, would become understandable when, many centuries later, Johannes Kepler developed his laws of planetary motion. With these adjustments, Ptolemy was able to construct a model which could give the positions of the celestial bodies, in the past and in the future, to about 1°, the accuracy of the observations. It was a very complex model and it required 55 spheres to carry the different celestial bodies, but he had achieved his life's aim of quantifying Aristotle's cosmology, and his model was to be used by astronomers (and astrologers) for a thousand years.

It should be stressed that Ptolemy's system allowed the *positions* of the heavenly bodies to be predicted; their *distances* were a quite different matter. Because of the work of Aristarchus, Eratosthenes and Hipparchus, the distance of the Moon was known accurately to Ptolemy, but the distances to the Sun, planets and stars were very uncertain, simply because those distances were unimaginably immense by terrestrial standards. The first good estimates of the distance to the Sun and the nearest stars had to wait until the eighteenth and nineteenth centuries respectively; now, towards the end of the twentieth, we are still struggling with the size of the Universe. Ptolemy adopted a neat plausible solution for the distances of the heavenly bodies, but without any observational basis. He supposed that the range in distance from the Earth of each of the heavenly bodies fitted exactly within the ranges of those on either side. In other words, the nearest approach of any planet would be at the farthest distance of the one inside it, and the farthest approach would be at the nearest distance of the planet outside it. In terms of his model, which had the planets revolving about the Earth in their deferents and epicycles, this made some sense, in that closer separations would risk collisions; but of course they could be farther apart. His model gave an order of distance for the heavenly bodies which placed the Moon as the closest and then, in order, Mercury, Venus, the Sun, Mars, Jupiter and Saturn; beyond the range of Saturn's epicycle, he placed the boundary of the Universe, the sphere of fixed stars. He then put the inner edge of Mercury's epicycle (the closest approach to Earth) at the upper edge of the lunar orbit; he then placed the lower boundary of the motion of Venus at the upper boundary for that of Mercury; and so on. Since he knew the distance to the Moon, he was able to calculate the distances of the prime celestial bodies. He concluded that the Sun was 20 times more distant than the Moon, whereas the actual value is about 400; for the planets he calculated that the most distant, Saturn, was nearly 300 times as far as the Moon, whereas the actual value approaches 4000. But of course his greatest error lay with the stars, which his scheme placed just beyond Saturn, and even the ten nearest stars that can be seen with the naked eye lie a million times farther than that at a distance which takes light from them an average of 12 years to reach us.

The numbers given above are of historic interest only because they give a feeling for the human view at that time of the scale of the Universe. Ptolemy's method of assigning (the author cannot say estimating) distances was completely invalid, but its neat and tidy concept, and his great distinction, made it generally accepted for more than a thousand years. It was strengthened by the fact that his distance to the Sun (20 times that to the Moon) agreed with that of Aristarchus, using a different but fully valid method; however, this lacked the precision of measurement required and the agreement was quite fortuitous. The question of distance in astronomy is fundamental and will recur repeatedly in this narrative as successive advances led to the realization of the immensity of the Universe.

Ptolemy either ignored the heliocentric views of Aristarchus or was unaware of them. He was very faithful to Aristotle but only insofar as the observations allowed. His deviations from Aristotelian cosmology were small in magnitude but important in principle. The Earth had been slightly displaced from the centre of the planetary orbits, and their velocities appeared constant only when viewed from another point (the equant), which was also displaced from the centre. Perhaps it was the fact that only *small* deviations from Aristotle were needed which caused Ptolemy to follow that route. Being an Alexandrian, he must have known of the heliocentric views of Aristarchus and he had the mathematical ability and the observations that could have developed a heliocentric model and established him as an early Copernicus. Whether it was the absence of parallax (attributable to the then unbelievable remoteness of the stars) or the mammoth philosophical strength of Aristotle, we shall never know. But more important for science, he had insisted on fitting the theory to the observations, thereby violating Aristotle's philosophy, derived from Plato, that logic and reasoning were the most effective ways of understanding the Universe, with no need for observation or experimentation.

Ptolemy marked the end of the great Greek period; indeed its heyday had been long past when he lived his life. The intellectual domination of Greece had been replaced by the military domination of Rome. Except for some detailed criticism and adjustments by Islamic astronomers in the tenth century AD, his *magnum opus*, the *Almagest*, was to endure as an astronomical bible for 14 centuries, when it was to be replaced, together with the natural philosophy and cosmology of Aristotle (which had endured five centuries longer), in the second great cultural and intellectual explosion of the human race – the Renaissance.

CHAPTER FOUR
The Interlude

Dark to me is the Earth. Dark to me are the Heavens.
. . . Desolate are the streets. Desolate is the city.
Blunt

After the great Greek period, scientific, and indeed all intellectual activities, declined in Europe and the Middle East. All of the Mediterranean countries and large parts of Europe were under the military domination of Rome for some six centuries and, by the time its empire collapsed in AD 476 (the eastern rump based on Byzantine, modern day Constantinople, lasted another thousand years), intellectual pursuits had effectively vanished in all areas of scholarship. The period that followed in Europe is now known as the Dark Ages; for 500 years intellectual and cultural endeavours ceased in all areas and for a further 500 years in others, which included science and astronomy. But these were revived in the Middle East, where Arabian mathematicians and astronomers led the world from the ninth to the thirteenth centuries.

No such decline occurred in the astronomical and cultural activities in the Indian and Chinese civilizations. They had developed calendars and, having measured the solar year, could keep them in phase with the seasons. Astronomical observations were conducted largely for astrological reasons; in China, astrology was based on unpredicted celestial changes, such as comets, eclipses, aurorae and new stars, all of which were supposed to herald major human events, usually catastrophic. Chinese observers were therefore attuned to such changes and, of particular note, led to their recording the appearance of the new star of AD 1054, enabling modern astronomers to accurately date the supernova that caused the Crab nebula, an object of intense recent study. Surprisingly, it was not recorded in the Western world, not even by the active Arabian astronomers, despite the fact that it was brighter than any of the stars or planets.

Astronomical studies were also greatly influenced by the religions of India and China. The earliest, Hinduism, which developed in India about 2000 BC, had neither a founder nor prophets, and it emphasized the best way of behaving rather than a set of doctrines. In the first millennium BC, Buddhism also developed in India and Confucianism and Taoism in China. All of these proposed a way of living in which the higher human qualities were stressed, such as benevolence, righteousness and filial duties. These became part of the social and cultural lives of the people of India and China, and can be regarded as philosophies as much as religions as is understood in the monotheistic faiths.

As was recounted earlier, religion also influenced the development of astronomy in Mesopotamia and Egypt, although its effect on Greek astronomy was minimal. In all the early great empires, polytheism had been the order of the day, with religions based on the existence of several gods, each with different responsibilities and functions, often vying with each other and having the prominent celestial bodies named after them. There was one small but notable exception which was ultimately to have a major effect on the development of world history. In the earliest Mesopotamian civilization of Sumeria, an idea germinated of a single, all-seeing, all-powerful God who created everything – life, the Earth, Sun, Moon, planets and stars – and whose will must be obeyed. This belief was adopted by a small semitic tribe of Hebrews (wanderers) under the leadership of Abraham; they were later also to be called Jews and their religion was to be known as Judaism. Abraham led his tribe out of Ur, the capital of Sumeria, in about 2100 BC, in search of the land promised by God in the Jewish covenant. Abraham was succeeded by his son Isaac, and Isaac by his younger son, Jacob, who was renamed Israel. This led the Hebrews to adopt the name of the People of Israel, and the 12 sons of Jacob (by his two wives and their two maids) formed the basis, according to Jewish tradition, of the 12 tribes of Israel between whom property had to be divided. The Hebrews were a nomadic people and were used to roaming through the extensive desert areas that were bordered by the "crescent of fertility", the strip of agricultural abundance running up from the Nile, through Palestine into Syria and down into the valleys of the Tigris and Euphrates. In about 1700 BC, they reached Egypt, where they were held in semi-bondage for several centuries before their exodus, under Moses, around 1200 BC. Joshua, who succeeded Moses, led the Israelites across the Jordan and into the promised land, Canaan (present-day Palestine). It was an efficient military operation in which the Canaanites were subdued (but not the southerly based Philistines), allowing the formation of the state of Israel. In about 1000 BC, the nation was formed into a Kingdom under its first King Saul, who was followed by David and then Solomon. This was one of the greatest periods in Jewish history when Israel was a single, identifiable, powerful nation. It was to be subjugated by Persia in about 500 BC and by the time of Christ, it was part of the Roman Empire. The Jews were to be dispersed throughout Europe and, until the second half of the twentieth century, were not to have their own state again. The factor that held them together and identified them as a people, was their religion.

In the post-Grecian period, two great new monotheistic religions developed which were to have major implications, not only for astronomy and science, but for all human activities. Christianity, which was based on the belief that Jesus Christ was the son of God (the same God as that of the Jews) was a breakaway from Judaism and it spread through the whole of Europe; it cannot be coincidental that its rise occurred in phase with the decline of the military power of Rome, which, by embracing the Christian religion and becoming its centre, saw its military power being replaced by a spiritual one. The second new monotheistic religion was Islam, which had the same ancient Sumerian roots as Judaism, the belief in a single all-powerful God. Its great prophet was Mohammed, who wrote the Koran on the basis of communications from God. The three monotheistic religions – in chronological order, Judaism, Christianity and Islam – therefore have the same common base, a belief in a single, all-powerful God whose will, or instructions, must be obeyed. God's will was communicated by different routes in each religion: in Judaism via several Jewish prophets, particularly Moses; in Christianity by the son of God, Jesus Christ, communicated by his disciples and the preaching of Paul of Tarsus; and in Islam, via the one and only prophet Mohammed. These routes resulted in different interpretations of God's will and, hence, despite their common root, they are not compatible in practice or in principle; for example, Judaism and Islam cannot accept the basic Christian belief that God had appeared in human form in the person of his son, Jesus Christ, because this was contrary to their concept of a remote, unseen, all-powerful God, who communicated only through chosen individuals, called prophets. Another major difference is that Judaism was, and still is, an inward-looking religion which protected the Israeli people from outside interference. Christianity and Islam, on the other hand, were very expansive and extended their influence through many countries, often backed by military force. Christianity spread through the whole of Europe, and Islam advanced through the Middle East as far as India, took over all of North Africa, and reached into Spain. Because of the power and extent of the Christian and Islamic faiths, their effect on human life was immense, but this story will be confined to the astronomical and philosophical developments that occurred in their areas of influence, both positive and negative.

Unlike other religions, the Christian church did not stimulate astronomical studies with religious requirements during the early centuries of its development. (In more recent centuries, an astronomical question has been posed and debated extensively, as to the nature of the star of Bethlehem quoted in the New Testament, something which has still not been fully resolved, although there are many proposals.) The belief was established that all truth lay in holy scripture as revealed by the gospels and, consequently, philosophical and astronomical studies, being unnecessary, were not encouraged. This formed the basis for a fundamental conflict between the Christian religion and science which was to continue for centuries. The conflict is well illustrated by the writings of one of the earliest of Christian theologians, Tertullian, who lived in about AD 160–220. He was a lawyer who was born in Carthage and visited Rome for a period during his thirties, where he was converted to Christianity. On returning to Carthage, he produced many major theses on the meaning and nature of Christianity and did so with the enthusiasm and conviction of the converted. He claimed that the scriptures were the source of all wisdom, and formulated the doctrine of the Holy Trinity – God the Father, God the Son and God the Holy Ghost – which he based on the Roman Triumvirate of Pompey, Caesar and Crassus, which was formed in 60 BC. His works were expressed in passionate and exaggerated language, and were the first to be written in Latin, which had become the most important language of the age, and they were to have a great influence on the development of Christianity and the Christian church. I have extracted a piece from one of Tertullian's writings which most effectively demonstrates the conflict that was to occur between the developing Christian church and the nascent world of science:

What has Athens to do with Jerusalem, the Academy with the Church?
. . . We have no need for curiosity since Jesus Christ, nor for inquiry since
the Gospel.

This is expressed in quite magnificent language in which the great ancient works of Athens and its Academy are taken to represent activities based on curiosity and inquiry – the basis for what we now call science. Of course the quotation is quite extreme in totally denouncing such motives, and perhaps overstatement was part of Tertullian's style, and he may sometimes have written with his tongue in his cheek. This possibility is supported by another quotation, also superbly written, in which he refers to the incident, reported by Plato, whereby the ancient Greek astronomer Thales was supposed to have fallen down a well while engrossed in observing the stars (mentioned earlier). Tertullian said:

Tell me, what is the sense of this itch for idle speculation? What does it
prove, this useless affectation of a fastidious curiosity, notwithstanding
the strong confidence of its assertions? It is highly appropriate that

Thales, while his eyes were roaming the heavens in astronomical observation, should have tumbled into a well. This mishap may well serve to illustrate the fate of all who occupy themselves with the stupidities of philosophy.

Tertullian is quoted only to illustrate the point that the Christian religion developed on the basis that the Gospel was the primary source of guidance and of truth, and was inviolate. This commitment to Holy Scripture was, and still is, the fundamental basis of Christianity, but there is no doubt that it was a discouragement to scientific endeavours and these languished for a thousand years after the military fall of Rome. During that time, possibly because the Gospel was based on ancient writings, other ancient works of a non-religious character, including the writings on science by the ancient Greeks, also became regarded as inviolate. These factors were to lead to one of the most unfortunate events in the history of Christianity and science – the trial of Galileo, which will be addressed later with the help of another quotation by Tertullian.

Islam developed in the sixth century AD and was essentially established by one individual, the prophet Mohammed, who wrote the Koran as the word of God (Allah) as divinely communicated to him. Like the Christian Bible, the Koran expressed the will of the Almighty and was therefore inviolate and could not be questioned. Unlike Christianity, the Muslim faith greatly stimulated scientific endeavours, mainly as a result of the heavy demands made by the religious requirements of Islam, which posed very difficult problems in mathematical astronomy.

On 16 July AD 622 (on our present-day Western calendar) Mohammed fled from Mecca to Medina, an event (the Hegira) which is taken to mark the initiation of Islam. The two cities of Mecca and Medina became the most holy cities, and the date of the Hegira was to become the first day of the first month of the first year in the Muslim calendar. After its establishment, in what is modern-day Saudi Arabia, Islam expanded rapidly and, by the end of the second Islamic century, extended through all of the Middle East, North Africa and most of Spain. At this point a more stable situation followed which saw the greatest mathematical and astronomical developments of the age (ninth to thirteenth centuries AD; second to sixth centuries Muslim) which were spurred by the demanding religious needs of Islam. It is fitting that these activities occurred in a region that embraced Mesopotamia and Egypt, where many of the earliest major achievements in astronomy were made.

In AD 762 (140 post-Hegira), Baghdad was founded as the capital of Islam, and it was there that the "House of Wisdom" was formed, which was to become a centre of learning like the great schools of ancient Greece. The decimal system of numbers, borrowed from India, was adopted and this Arabic system became the basis for future numerology; other seminal mathematical developments were made in spherical geometry and trigonometry. The great Greek scientific and philosophical writings were translated extensively into Arabic and, as in the Christian world, were accorded the status of truth, which was bestowed on antiquity by both religions. Islamic astronomy was so dominant that many astronomical words in use today are of Arabic rather than Grecian origin. These include star names such as Betelgeuse, Rigel, Vega and Deneb, and technical terms such as the zenith; in addition, Ptolemy's great work, entitled the *Syntaxis* in Greek, is now better known by its Arabic name, the *Almagest* (the Greatest).

One of the challenges to astronomers posed by Islam was due to its calendar. In Mohammed's age, the other earlier monotheistic religions of Judaism and Christianity determined the times of their great festivals, such as the Passover and Easter, from the phases of the Moon, because they were dated from the early Jewish calendar, which was lunar-based. This resulted in the number of months in a year being about 12 .33, which, not being a round number, required cumbersome adjustments to keep the seasons in step. This offended Mohammed, who decreed in the Koran that, in the sight of God who created everything, there were 12 months in a year. In setting up the Muslim calendar, this was interpreted, understandably, as meaning 12 months as defined by the Moon. Hence, the Muslim calendar year is about 11 days shorter than a solar year, causing the seasons to migrate through the calendar so that they complete a full cycle about every 33 years. Of course, the Arabian astronomers were fully aware of this and were able to date the seasons within their calendar during any year, but festivals were a different matter and had to be held on the same calendar dates in every calendar year. Since the first day of each month began with the appearance of the new Moon in the early evening, its start was uncertain by one day since it would occur on either the 29th or 30th day of the preceding month, and its visual detection, which had to be corroborated by at least two observers, could be affected by subjective factors and climatic conditions. This uncertainty did present a handicap to the detailed preparations for the great religious festivals such as the holy month of Ramadan. This is the ninth month of the Muslim calendar and is a period of strict fasting between dawn and dusk, so knowing precisely which day it would start on before the event would be a distinct advantage in preparing for it. This posed a challenge to the Islamic astronomers, since it required the Moon's position to be predicted accurately with respect to the local horizon, a very difficult problem in spherical geometry which they tackled successfully enough for the purpose.

The problems of the Islamic calendar were handled by many astronomers without any one of them standing out particularly above the others. However, one individual is worthy of mention, not because of his contributions to the calendar, significant though they were, but because, unlike the other Islamic names given in this chapter, his name will be familiar to most readers. Omar Khayyam was a Persian mathematician and astronomer who was born in the middle of the eleventh century AD and lived on into the twelfth century. He wrote a book on algebra, which established him as one of the leading mathematicians of the time and caused the Sultan to summon him in 1074 and instruct him to reform the calendar thoroughly. But that is not the reason why he is known so well today; his fame stems from his other talent and hobby, the writing of poetry, in which he demonstrated the insight and understanding of a keen independent mind. He composed his verses in the quatrain form – stanzas of four lines with some form of rhyming – and in many such verses, he presented his views and philosophy on the questions of life and death. These were discovered in the nineteenth century by the English poet, Edward Fitzgerald, who translated them, very freely, to produce *The Rubaiyat of Omar Khayyam,* one of the most famous and quoted poems in the English language. It has 75 verses, or Rubaiyat, and Fitzgerald maintained the quatrain stanza form in which the first, second and fourth line are rhymed. Although the original thoughts are those of Omar Khayyam, these were expanded by Fitzgerald, and the language and poetry that have made the work so lastingly popular are entirely attributable to him.

In addition to the calendar, two other problems needing astronomical solutions were posed by the customs of Islam. The first was the requirement to pray towards Mecca, whose direction therefore had to be determined from any part of an extensive empire ranging from Spain to Persia; the second was the need to pray at sunrise, midday, afternoon, sunset and evening, requiring the determination of local time through that same vast empire. Direction was determined from the positions of the Sun and stars, and time from the direction and altitude of the Sun. Neither of these methods was completely new, but the Arabian scientists made them much more rigorous with the aid of their seminal developments in spherical geometry and trigonometry.

The great achievements of Islamic astronomy were theoretical rather than observational in nature, primarily because Arabic astrology depended almost entirely on the positions of the planets, which could be predicted well enough from Ptolemy's *Almagest.* There was therefore little astrological stimulus to making astronomical observations, and observatories were built mainly for astronomical reasons, such as those needed to improve Ptolemy's world model or for time and calendar determinations. A major observational work appeared in AD 963 when Al Sufi

47

published his *Book of Fixed Stars*; this updated Ptolemy's catalogue of the second century AD, which was itself mainly based on the observations of Hipparchus made in the third century BC. In his book, Al Sufi reports the existence of a faint extended object present in the constellation of Andromeda in the celestial sphere of fixed stars. This is the earliest known record of an astronomical nebula and we now know it to be the great spiral nebula of Andromeda, whose study in the early twentieth century was to make a dramatic contribution to our knowledge of the scale and nature of the Universe.

The cosmology of Islam was guided mainly by Aristotle and Ptolemy, whose writings had been translated into Arabic and studied in depth. As related earlier, Ptolemy had been faithful to Aristotle, departing only slightly from his theses when the need to match the observations of Hipparchus demanded. The Muslim astronomers were to be even more faithful to Aristotle, whose statements on cosmology they accepted totally and without question. From the eleventh to the thirteenth centuries, they started to question and then criticize Ptolemy for the small adjustments he had made. Before that period the Ptolemaic approach was accepted fully and efforts were concentrated on deriving better values of the basic parameters in order to make the calculations more accurate. In the ninth century, a brilliant astronomer, Muhammad al-Battani, significantly improved the Ptolemaic model, mainly by determining the orbit of the Sun against the fixed stars much more accurately than Ptolemy had managed. This enabled him to construct his own astronomical tables, which formed the greatest astronomical work since the *Almagest* and which they were ultimately to replace. Via Spain, the periphery of the Muslim empire, a Latin translation reached mediaeval Europe, where it was printed extensively and used by Copernicus, who referenced it several times in his great work *On the Revolutions of the Heavenly Bodies*.

Criticism of Ptolemy started in the eleventh century, when a leading philosopher, Ibn al-Haytham (based in Cairo) wrote a book called *Doubts on Ptolemy*. This was followed in the next century by an even greater criticism by Ibn Rushd of Andalucia, then part of Islam. Their problem was caused by the slight adjustments from the strict interpretation of Aristotle that were made by Ptolemy in order to fit the observations. Aristotle had placed a stationary Earth at the centre of the Universe, with its outer bound, the sphere of fixed stars, revolving around it at a constant speed; all other celestial bodies were also carried on spheres centred on the Earth and rotated at constant velocity. As explained previously, Ptolemy had used the ingenious concept of the epicycle to explain planetary motions; this was a circle around which the planet moved at constant speed but whose centre moved at a constant (but different) speed around a larger circle (the deferent) which had the Earth as its centre. This agreed with

Aristotle: motions were circular at constant speed and the stationary Earth was at the centre. But Ptolemy could not match the observations exactly, so he made two slight adjustments that did allow a match. He placed the Earth just off the geometric centre of his system and he displaced the dynamic centre (the equant about which the velocities were constant) by the same amount but on the other side. These adjustments were numerically small, but in the eyes of the Islamic astronomers of the eleventh to thirteenth centuries, they were large in principle. They also offended the mechanical concept of a rigid rotating sphere carrying the celestial bodies, for which the geometric and dynamical centres had to be the same. These doubts were expressed by Ibn al-Haytham and were greatly strengthened by Ibn Rushd, who proposed that there should be a strictly geocentric Universe with all bodies carried by strictly concentric spheres.

This point of view took the problem back to the stage that Ptolemy reached when he could not quite fit the data with his deferents and epicycles. He had taken his eccentric sphere route, but this was now discredited in mediaeval Islam, and astronomers embarked on the very difficult problem of building a model based entirely on concentric spheres. This was finally solved in the middle of the fourteenth century by Ibn al-Shatir, an astronomer working in Damascus. Using an earlier idea, he introduced the secondary epicycle, an additional circle around which the planet moved at constant velocity and whose centre moved at constant velocity around the epicycle, the centre of which moved in turn around the deferent, whose centre was the Earth. This use of deferent, epicycle and secondary epicycle allowed a truly geocentric system with strictly concentric spheres, which explained the observations of the Moon, Sun, planets and stars. Aristotle's cosmology had finally been quantified, but this work did not reach Europe, unlike the astronomical tables of al-Battani, and therefore had no influence on the major astronomical developments that were to occur there a few hundred years later.

During the great period of Islamic astronomy, scientific endeavours and inquiry were virtually absent in Christian Europe, largely because of the lack of encouragement (some would say active discouragement) for such endeavours by the Christian Church, which believed that truth and spiritual guidance could come only from holy scripture and that natural knowledge and understanding was best revealed by ancient writings. Parallel views were held by Islam, but that religion, unlike Christianity, greatly stimulated scientific activities, mainly mathematical and astronomical, during the first several centuries after its establishment. It is therefore paradoxical (the perceptive historian may say understandable) that the really great scientific revolution that was to come and which was to initiate and establish modern science, with all its vast implications for the development of the human race, was to occur in the heart of Christendom.

CHAPTER FIVE

The Renaissance

To realise that our knowledge is ignorance,
This is a noble insight.
To regard our ignorance as knowledge,
This is mental sickness.
Tao Teh Ching

The Renaissance is one of the truly great intellectual and cultural periods in human history. It was mainly centred on Italy but spread through the whole of Christian Europe. The word means "rebirth" and it was used to mark the resurgence of creative activities in Europe after the long interval of the Dark Ages. It is always difficult to give a date to some period of history defined in a particular way, but historians usually assign the Renaissance to the three centuries starting with the fourteenth. This was a century during which Europe faced peril on an unprecedented scale. A fatal, virulent, bubonic plague, called the Black Death, swept through Europe from Asia in the middle of the century and killed about 25 million people, nearly a third of the population. Subsequent epidemics during the rest of the century greatly added to the death-roll; never had dying been witnessed on such a massive scale, dwarfing even the worst of human conflicts. It therefore seems remarkable that it was this century that saw such a great outburst of artistic activities and achievements marking the start of the Renaissance.

But there were signs of the development of human creativity in Europe before the fourteenth century, starting in the eleventh. These were stimulated by the Christian Church, which, as has already been said, did not greatly encourage the pursuit of scientific or philosophical activities, but it did encourage the pursuit of another area demanding human creativity – architecture – for the purpose of designing and building cathedrals. These were the palaces in which the bishops, among the most powerful dignitaries in Europe, presided, and the results were accordingly marvellous demonstrations of human art and ingenuity. One example is the magnificent cathedral of Notre Dame in Chartres, France, which lies on the left bank of the River Eure some 80 kilometres southwest of Paris. On a site of an earlier church, the great cathedral was built in the twelfth century and is one of the most beautiful examples of Gothic architecture. These activities were a preface to the Renaissance proper, which, in the fourteenth to seventeenth centuries, saw an explosive re-emergence of the human creative force in all areas of scholarship, in art, in sculpture, in literature, in music and in science. The great intellectual and cultural achievements of Christian Europe during the Renaissance were such that they not only equalled but even surpassed those of ancient Greece.

The great scientific achievements of the Renaissance were mainly of an astronomical nature. They were very overdue, because in the 14 centuries that followed the publication of Ptolemy's world model, the *Almagest*, progress in astronomy (and science) was minimal. Although observatories were built and the positions of stars and planets were a matter for study, particularly in the Arab world, the understanding of the known Universe did not advance one iota. Even the great Islamic astronomers concentrated their efforts, beyond those needed to meet the requirements of the Koran, in improving and consolidating Grecian astronomy. It must be one of the great mysteries of history to explain why the human race, endowed with such intelligence, seemed to lose its curiosity about natural phenomena for so long and was willing to accept, without question, the writings handed down from ancient Greece. These writings represented a tremendous legacy in philosophy and scholarship, but in science much of the good (in mathematics and geometry) was counteracted by the dogma of Aristotle in cosmology and natural philosophy, which presented an impediment to the development of science through the Dark Ages and into the Renaissance. The scientific revolution that was now to occur laid the foundation of modern science; it started in the sixteenth century in the later stages of the Renaissance and was complete before the end of the seventeenth century. By then, science had been placed on a firm, rigorous basis which, unlike the post-Grecian era, was to continue to flourish and thrive to the present-day causing, en route, technological developments that have revolutionized the quality of human life.

The period we are now considering, Christian Europe in the sixteenth century, was totally different from that of ancient Greece in its heyday some 2000 years earlier. It was different in all aspects, in customs, in dress, in politics and religion; but in one area, cosmology, little had changed: Aristotle still reigned supreme, although his spheres carrying the planets had been replaced by angels pushing them. The story of the great scientific revolution of the Renaissance can be told in the achievements of five individuals who were of different nationalities, had different origins, possessed different temperaments, different personalities and different abilities, but all having one thing in common – a touch of genius. They were Nicolaus Copernicus (a Pole), Tycho Brahe (a Dane), Johannes Kepler (a German), Galileo Galilei (an Italian), and Isaac Newton (an Englishman). They were to uproot the Aristotelian precepts in natural philosophy and cosmology completely, and replace them with the new science.

Nicolaus Copernicus was born in 1473 in the old Hansa town of Torun on the banks of the Vistula in Poland. His father was a merchant who died when Nicolaus was only ten years old and he was looked after and guided by a benevolent uncle who was to become a bishop. He was sent to school in Torun, after which he entered the renowned Polish University of Cracow, where his studies included mathematics and astronomy. At the age of 23, he went to Italy, where he studied the liberal arts, medicine and canon law at the universities of Bologna, Padua and Ferrara. He was fortunate in receiving such a wide and extensive education, and he returned to Poland after seven years in 1503 where he entered the Church as a canon of the cathedral of Frauenberg, in which capacity he was to serve the Church with devotion for the rest of his life. There is little doubt that his career in the Church was productive and of high standing; but, distinguished as it was, he would not be present in this story, nor would his name be known in history, if it were not for his great passion and hobby, astronomy, to which he devoted considerable amounts of his time. Like the great astronomers of antiquity, his fascination lay in understanding the motion of the planets and, having available to him translations of Ptolemy's *Almagest* of the second century AD and its improvement by the astronomical tables of al-Battani in the ninth century, he studied these deeply. But Ptolemy's world model offended him by its complexity and he had an inner belief that nature must be basically simple, a belief held by many of the great philosophers.

Copernicus therefore embarked on a new world model, based on the proposition that the Sun, not the Earth, was the centre of the Universe, and this did indeed produce a much simpler system. As recounted earlier, the brilliant Grecian astronomer, Aristarchus, had the same idea some 18 centuries before, but we only know of this indirectly, via the writings of Archimedes, which reported it as an idea and did not give any quantitative calculations of the precept. Whether Copernicus knew about this is still not completely certain, but it probably does not matter. I am sure that Copernicus reached the concept independently, without historical stimulation, and, more importantly, he fully developed it himself, thereby producing a new quantitative heliocentric cosmology which was to have a major impact on human thought.

Copernicus placed a stationary Sun at the centre of the sphere of fixed stars, which was also stationary. The planets revolved around the Sun in the same plane (the ecliptic), moving in circles at constant speeds. Their order of distance from the Sun was Mercury, Venus, Earth, Mars, Jupiter and Saturn. The Moon, however, revolved around the Earth but in the same ecliptic plane, and the Earth had two motions: an orbital revolution about the Sun every year and a rotation about its own axis every 24 hours. Its axis was inclined to the plane of its orbit, which, as described earlier, gave a simple explanation of the annual change in the seasons. In his calculations, Copernicus first had to transform the positions of the planets, observed from Earth, to positions relative to the Sun, but he was then able to draw on the mathematical techniques developed by Ptolemy and the Islamic astronomers in order to finalize his calculations. The result was a model that was immensely simpler than that of Ptolemy, and one that could be easily envisaged and understood. But the concept of the Earth orbiting around the Sun every year at an immense speed and rotating about its axis every 24 hours, posed a great mental obstacle to its acceptance, and another obstacle was to be added of a religious nature. Since his world model was to become the subject of bitter debate and dispute, I will comment on it scientifically, but in the framework of the time. There were no astronomical observations that specifically favoured the Copernican system over that of Aristotle and Ptolemy. The heliocentric model made the specific prediction of stellar parallax – the apparent wobble in the position of a star as the Earth moved from one side of its orbit to the other side – whereas the geocentric model predicted none, and, indeed, none had been detected. We now know that this is because of the immense distance to even the nearest stars, which makes their parallax so small that it was not detected until three centuries later when telescopic techniques were developed sufficiently. But at the time, the only argument in favour of the heliocentric system was an aesthetic one – it had great simplicity and form – yet this was sufficient to convince the most important scientific minds who were to follow Copernicus in the Renaissance period.

The only, albeit major, change in the Copernican system, compared to that of Ptolemy, was placing the Sun at the centre of the Universe rather than the Earth; in all other ways Copernicus was true to Aristotle, with the celestial bodies, including the Earth, moving in circles at constant speed. But, like Ptolemy, he found he could not get a best fit with the observations and, like Ptolemy, he had to make small corrections and he chose these to be the same ones as developed by Ptolemy but adjusted to the heliocentric system. His model immediately explained the greatest problem faced by a geocentric system – the retrograde motion of the planets – and therefore did not need the large epicycles used by Ptolemy; but, in order to fit the observations, Copernicus used small epicycles and also placed the Sun at a point slightly displaced from the geometric centres of the planetary orbits. The reason why these adjustments were necessary was to be explained later by Kepler's first two laws of planetary motion.

Copernicus held up publication of his momentous work for nearly 20 years until the year of his death in 1543, when his book *De Revolutionibus Orbiun Coelestium* (*The Revolutions Of The Heavenly Orbs*) appeared. Why he delayed is a matter for speculation and history, but perhaps his insight into astronomy extended to his perception of the attitudes of individuals and institutions, allowing him to foresee the possible conflict his proposal might cause in the Church, a conflict that eventually flared up in the trial of Galileo. He was a contemporary of Leonardo da Vinci, who was 21 years his senior, and of Michelangelo, who was two years his junior, and who sculpted his *David* when Copernicus was 30 and completed painting the Sistene Chapel ceiling when Copernicus was 41. He was therefore highly fortunate to live at a time when some of the greatest works of human art were being created, but it was also a time that witnessed the start of religious conflict and division within Christian Europe. Copernicus was a contemporary of Martin Luther and, at the age of 44, he was probably deeply engrossed in the calculations needed to construct his heliocentric model when, in 1517, Luther nailed his 95 theses to the door of the church in Wittenberg on the River Elbe. Copernicus would therefore have been fully aware of the momentous events occurring in neighbouring Germany, where the established Church, of which he was a loyal canon, was being challenged in the early stages of the Reformation, which was to result in the Protestant breakaway from the Roman Catholic Church.

The historic and scientific legacy left by Copernicus lies entirely in one activity, the construction of his heliocentric world model. This makes me think of a couplet from a Shakespearean sonnet, which seems particularly appropriate:

Then how when Nature calls thee to be gone,
What acceptable audit canst thou leave?

On behalf of Copernicus, the answer could be given by simply saying "one book", but what a book! When he died in Frauenberg, East Prussia (now Frombork, Poland) Copernicus could have had no idea as to the impact his book was to have on the development of science and philosophy. He was not to know that the giant intellectual step he had taken was to have such a traumatic effect on human thought that the one word in his title (which simply meant going around) by which his book became known was to acquire a new and totally different meaning which has virtually superseded the old: revolution.

The Revolutions was a very poorly written book to the extent that it

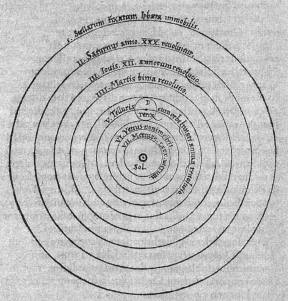

NICOLAI COPERNICI

net,in quo terram cum orbe lunari tanquam epicyclo contineri
diximus. Quinto loco Venus nono menfe reducitur. Sextum
deniqʒ locum Mercurius tenet,octuaginta dierum fpacio circū
currens.In medio uero omnium refidet Sol. Quis enim in hoc

pulcherimo templo lampadem hanc in alio uel meliori loco po
neret,quàm unde totum fimul pofsit illuminareⵈSiquidem non
inepte quidam lucernam mundi,alij mentem, alij rectorem uo=
cant. Trimegiftus uifibilem Deum,Sophoclis Electra intuentē
omnia,Ita profecto tanquam in folio regali Sol refidens circum
agentem gubernat Aftrorum familiam. Tellus quoqʒ minime
fraudatur lunari minifterio ,fed ut Ariftoteles de animalibus
ait,maximā Luna cū terra cognatione habet.Concipit interea à
Sole terra , & impregnatur annuo partu. Inuenimus igitur fub
hac

The Copernican world model taken from the first edition of his memorable *De
Revolutionibus* and showing the Sun at the centre with the planets, including the
Earth, orbiting about it but with the Moon orbiting the Earth.

could only be read and understood by a few experts whose summaries of the work became the main method by which its doctrine was spread. One part of *The Revolutions*, the preface, was not written by Copernicus but by Andreas Osiander, a leading Protestant theologian and one of the early builders of the Lutheran creed. Osiander handled the publication of the work and wrote a preface which was inserted anonymously and was therefore believed to be attributable to the author himself. It stated that the contents of the book and the theory it propounded should be treated as hypothetical and regarded only as an intellectual and mathematical exercise that was not intended to represent physical reality. It then went on to give arguments against the central thesis and implied that some of its consequences were absurd. These latter statements suggest that Copernicus never saw the preface and that it was inserted without his knowledge, and if so, it represents a somewhat underhand act. But the purpose in describing it here is not to debate whether the author saw it or did not see it, but to use it as an illustration of the dispute, more religious than scientific, that was to arise as a result of the Copernican doctrine and to be dramatically embodied in the person of Galileo. The preface, written by a leading figure in the emerging Protestant faith, was to become the official position of the Roman Catholic Church in its attitude to the heliocentric theory of Copernicus. This was that it could be discussed as an abstract intellectual exercise as long as it was not argued that it was factually correct. In this respect, the Catholic Church and the breakaway Protestant church, so severely in conflict at that time, appear to have been in agreement.

In carrying out his calculations of the heliocentric model, Copernicus had used the ancient observations of Hipparchus (with some additions by Ptolemy and improvements by the Islamic astronomers) made some 17 centuries earlier, since, remarkably, they were still the best available. This situation was to change dramatically to the extent that, by the end of the century that saw the publication of *The Revolutions*, a new, consistent set of astronomical observations was available of such quality that they completely superseded all previous data. Over a period of about 30 years, the positions of the stars and planets were measured with an unprecedented accuracy, one minute of arc, some 50 times better than achieved by Hipparchus. The architect of this programme was a Danish nobleman, Tycho Brahe, the greatest observational astronomer of his time and the greatest since Hipparchus.

Tycho Brahe, who is always known by his first name, was born in 1546 at the family seat in the southern tip of Sweden, which was then under the

Danish Crown. He was the eldest son in a famous and wealthy noble family which had close connections with the Crown, and he was brought up in the aristocratic society of Hamlet; indeed his father was governor of Helsingborg Castle, now in southern Sweden, which faces Elsinore, the scene of Shakespeare's play, across the sound on the island of Zealand in Denmark. While still a child, Tycho was adopted by a rich childless uncle, a vice-admiral who sent him at the age of 13 to study philosophy at the University of Copenhagen. After three years, the vice-admiral considered that further education for Tycho should be at a foreign university and sent him to study law at Leipzig, whence he proceeded to continue his studies in Wittenberg, Rostock, Basle and Augsburg, all German universities, with the exception of Basle in Switzerland. While at Rostock, he fought a duel with another young Danish aristocrat over an argument stemming from conceit as to their mathematical abilities, in which his nose was sliced off. From then on he wore a false nose made out of gold and silver. In 1574, two years after his return to Denmark, he married a peasant girl, to the surprise and dismay of his family and friends.

After this extensive education, in which his official studies were greatly neglected because of his passion for and studies of astronomy, Tycho returned to his native Denmark at the age of 26. He returned as a man wealthy in his own right because, during his studies, his uncle died prematurely, leaving a fortune to Tycho. The vice-admiral had returned to Denmark after a successful naval battle and was in procession with the king when, while crossing a bridge, his majesty accidentally fell over into the water, whereupon Tycho's uncle immediately dived in and rescued the king from drowning. However, this glorious act had a tragic conclusion, because he caught severe pneumonia and died shortly afterwards, with Tycho the beneficiary.

Tycho's interest in astronomy was an early development. While in Copenhagen, at the age of 14, he witnessed a partial eclipse of the Sun and this fired a fascination for astronomy, which never left him. When in Leipzig, he devoted much of his time to astronomy rather than law, the subject of his studies, and purchased an astrolabe with which he made his first astronomical observations, one of which, made at the age of 17, was of a conjunction of Jupiter and Saturn. These two events, the eclipse and the conjunction, fascinated him and he was very impressed by the fact that astronomers had predicted them, using Ptolemy and subsequent tables. But he was concerned by the inaccuracy involved, which was a few days in the case of the conjunction, and he therefore decided to embark on a career of astronomical observations with the specific aim of repeating the ancient observations of the positions of the stars and planets, but with greatly increased accuracy in order to allow more accurate predictions of future astronomical events.

Coming from a wealthy family and being wealthy himself, he was able, on returning to Denmark from his studies, to pursue his interest in astronomy energetically and to build a small observatory for that express purpose. Then, not long after his return, a remarkable astronomical event occurred on 11 November 1572: a new star, brighter even than Venus, appeared in the constellation of Cassiopeia, causing widespread fascination throughout Europe. Tycho measured its position regularly over 18 months before it became too faint for detection by the human eye. Its position in the celestial sphere did not change, it was exactly fixed relative to all the other stars; it really was a new star which had appeared suddenly and slowly faded. Tycho published his findings and these not only attracted great public attention but also contradicted Aristotle, for whom the celestial sphere was fixed and unchanging; any changes could only occur within the local Earth–Moon system. But Tycho had demonstrated fully that it was a new star in the celestial sphere and not a local object. We now know that the event that Tycho observed was a supernova, the first to be seen since 1054 by Chinese astronomers, and was the result of a catastrophic explosion in the final stages of the evolution of a massive star. The reasons for such remarkable events will be discussed and explained later in the context of stellar evolution.

Tycho's studies of the supernova brought him great fame at the relatively young age of 28. This led the King of Denmark, Frederick II, to offer his patronage on a very generous scale. The island of Hven, near Copenhagen in the sound between Denmark and Sweden, was made fully available to Tycho, together with the funds and costs to build and operate a major astronomical observatory. Tycho selected his site, the highest point on the island, defined the buildings and instruments, and work started in 1476. The observatory was to be called Uraniborg – Castle of the Heavens – after Urania, the muse of astronomy in Greek mythology. It was to have buildings for living as well as observing, and it contained a full suite of instruments designed by Tycho, with a full complement of support staff; it was the forerunner of the great modern observatories and would take several years to construct. Tycho supervised everything, from the buildings to the instruments, and did so in his characteristically extravagant style with little regard for cost, resulting in lavish buildings but also superb instruments. He designed his instruments with meticulous precision, in the perfectionist style that he always followed, and the result was an observatory with the finest and most accurate suite of astronomical instruments ever assembled. In these, he used the same concept of measurement as that of Hipparchus: an arm which could be rotated about some fixed axis, pointed towards the object under observation by visual line of sight, and its angle to the vertical or horizontal then measured. Tycho had spent much time at his several universities developing and perfecting such quadrants, and he was now able to push the concept to the limit, building huge high-precision quadrants in brass and oak, some nearly 12 metres in diameter and requiring four handles for their operation. In this way, he removed those errors caused by imprecision in the measuring equipment which had dominated the ancient observations, and he was left with only the limiting accuracy of the human eye – one minute of arc or one-sixtieth of a degree, a remarkable achievement for the time and a dramatic and astronomically significant improvement on the one degree that had been available previously.

Tycho seems to have been blessed by striking astronomical events which occurred at very propitious times in his career. The first was the supernova of 1572 and, one year after the initiation of the construction of Uraniborg, another celestial event occurred which attracted great widespread attention. This was the great comet of 1577. Although the observatory was far from complete, he did have one of his smaller quadrants operational and was able to observe the comet and measure its position as it moved through the sky. The comet was very bright and easily visible; indeed, like Venus, it could be seen before sunset, and had a tail stretching over 20° in the sky. First sighted in November in the constellation of Sagittarius, it moved swiftly and, when it was last seen in January 1578, it had reached the constellation of Pegasus. Tycho observed its position and structure over the ten weeks that it was visible, but did not publish his data for some time. He waited, with considerable foresight, until the observations made by others throughout Europe had been published. In common with the traditional historical view of comets as an omen of some imminent disaster, these reports discussed the possible nature of the impending tragedy, but Tycho had no interest in these speculations and he extracted what he was looking for – the recorded positions of the comet. He noticed that all observations, including his own, placed the comet in the same position at the same time against the background of stars, even though they were made from different and well spaced locations in Europe. They did not have the same precision as he had, but the absence of parallax was sufficient to deduce that the comet could not be an atmospheric phenomenon as was generally supposed. His analysis and consideration of the errors involved led him to conclude that the comet was more distant than the Moon. He published his findings in 1588 and his paper, devoid of astrology, took another chip out of the Aristotelian legacy, as had his studies of the supernova of 1572. Aristotle had stated that all observed celestial phenomena which demonstrated change, such as rainbows, meteors and comets, had to belong to the sublunar world, but Tycho had shown clearly that the great comet of 1577 was farther away than the Moon and was probably as far away as the planets.

On completion of Uraniborg, Tycho embarked on an observing

programme that was the best-organized, most comprehensive, most consistent, most continuous and, of greatest importance, the most accurate ever conducted. The positions of 777 fixed stars and of the 5 planets as they moved over the celestial sphere were recorded over 20 years to an accuracy of one minute of arc, thereby creating the most extensive and precise set of astronomical data ever assembled. They were to be superseded only by the later invention and development of the telescope. Tycho had achieved fame with his studies of the supernova and the great comet, but those studies, exciting and important as they were, were dwarfed as a contribution to astronomy by his stellar and planetary observations, which were to be the basis for the greatest advance in the then human understanding of the Universe since the ancient Greeks. It represented his life's work and he regarded it, justifiably, as his personal treasure.

The termination of Tycho's observing programme was not planned but was the result of the death in 1588 of his patron, the old King Frederick, who was succeeded by his son, Christian. Christian knew Tycho, having made enjoyable visits to Uraniborg in his youth, but he did not have the easy-going forgiving personality of his father. Tycho had become even more vain and arrogant during his time on Hven, was treating his tenants very badly on the island, and had increased his indulgence in private life by hosting many sumptuous banquets. He expressed no gratitude whatsoever for the generous (many would say overgenerous) patronage of the Danish Crown and made the mistake of not according the new king with the deference expected for a monarch. He wrote an impertinent letter to Christian, demanding additional support and implying that, if this was not forthcoming, he might seek the support of other kings or princes and leave Denmark. The new king responded with a firm letter which conceded nothing to Tycho and, indeed, stated plainly that continued patronage was entirely dependent on a major change in Tycho's behaviour and attitude to the monarch, with a clear demonstration of his appreciation and gratitude for its patronage. Christian's attitude was probably made more resolute by the fact that Tycho's observatory, Uraniborg, was a significant drain on the royal purse. The tone of the letter was such that Tycho realized that he had made a major error, that he had lost the royal patronage and that he would have to find another sponsor if he was to continue his work. It took him two years to find one, but he was finally successful and in 1599 he moved to Prague as Imperial Mathematicus to the court of the Holy Roman Emperor, Rudolph II, at the staggering annual salary of 3000 florins (the average salary of a professional person at that time was about 200 florins). The emperor gave Tycho the castle of Benatek, 35 kilometres northeast of Prague, together with funds to build another great observatory to which Tycho's famous instruments from Uraniborg were to be transported and installed.

In the two years leading to his arrival in Prague, Tycho had corresponded with a young German mathematician, who was working in the
provincial school of Graz in Austria. The young man had written his first
letter to Tycho in order to outline an orderly, mathematical scheme of the
Solar System based on the Copernican heliocentric model. Tycho knew of
and had studied the Copernican doctrine but could not accept it, for one
very good reason. The heliocentric system made one specific prediction
which, if confirmed, would decide unambiguously in its favour. As the
Earth moved around its orbit, the stars would appear to wobble as they
were viewed from a slightly different angle. This stellar parallax had never
been detected in the past and it also defied Tycho, even with the greatly
increased accuracy of his observations. As a great observer, he had a feel
for errors of measurement and he calculated that, if Copernicus was right,
non-detection of parallax would require the stars to be at least ten thousand times farther away than the Sun, a distance so great that he could
not accept it. (We now know that the nearest of the stars visible to the
human eye are about a million times farther away than the Sun). Tycho
also found it difficult to accept the Ptolemaic system fully, because of the
apparent artificiality of the planetary motions as represented by epicycles
centred on circles and not on a celestial body. He therefore developed his
own world model, which was intermediate between Copernicus and
Ptolemy, but closer to the latter. It was a geocentric model, with a stationary Earth at the centre, the celestial sphere of stars rotating about it, as
did the Moon and Sun, all in circles at constant speed. So far, this is the
same as Ptolemy, but the one change was that the five planets (not the
Earth) revolved around the Sun, also in circles at constant speed. With
this in mind, Tycho responded to the young mathematician, giving him
encouragement and making the suggestion that the concept be applied to
his, Tycho's cosmology, rather than that of Copernicus. It was clear that
he was impressed by the young man (rightly as it turned out) and he
invited him to join him in Prague. There were many reasons for this young
German mathematician to leave Graz in order to join Tycho in Prague, but
one was dominant over all the others; he was after Tycho's treasure. His
name was Johannes Kepler.

Kepler was born in Weil-der-Stadt, near Württemberg in southwest Germany, at 2.30pm on 27 December 1571. The date and time were recorded
later in life by Kepler himself, with that degree of precision that was to
mark his life's work. He was the eldest of seven children, three of whom
died in childhood, and Johannes himself was a very sickly child who was
smitten by smallpox at the age of six and suffered ill health throughout his

life, a condition exacerbated by very poor eyesight and a touch of hypo-chondria. He was born into a very poor Protestant family whose father, without any trade or craft, became a mercenary and went off to fight the Protestant insurgents in the Netherlands, an act that caused the family to be ostracized in the very Protestant town of Weil. It was the first but not the last event caused by Christianity's religious conflict that was to affect the life of Johannes Kepler. His childhood was not a happy one, with inces-sant arguments occurring in a small house between mother, father (when he was there), children and the grandmother who lived with them.

Kepler's family was so poor that, had it lived in most other parts of Europe, he would not have received any education, and he was very for-tunate (as was astronomy) that a tolerant and munificent mode of educa-tion had been introduced by the Dukes of Württemberg, who had embraced the Lutherian doctrines. They set up a system of grants and scholarships for poor and gifted children from God-fearing Christian homes committed to the true Protestant faith, with the aim of ensuring a supply of bright, fully educated students who could enter the priesthood and defend the Protestant Church in the major religious arguments sweeping through Europe. Kepler's obvious brilliance won him the necessary scholarships which allowed him an education at school, seminary and at university, the great Protestant centre of learning, the University of Tübingen.

He was first sent to a local elementary school, but his ill health, and the periods when he was sent out to work in order to earn some extra money, made his attendance very irregular and it took him three times longer than normal to complete his studies, which he did not finish until the age of 13. After four years at a theological seminary, he went to Tübin-gen, where he graduated from the Faculty of Arts when 20. Being destined for a priesthood in the Protestant Church, he entered the Theological Fac-ulty and studied divinity for another four years before an event happened that ended his religious studies and was to alter the course of his life totally.

In southeast Austria the province of Styria was ruled by a Hapsburg prince who belonged to the Roman Catholic Church, but the province also had a significant Protestant constituency. Reflecting this division within the Christian religion, the two centres of learning and teaching in its capital, Graz, were also divided on those lines, the university being Catholic and the provincial school Protestant. The latter institution, on the death of its mathematical teacher, approached the Protestant University of Tübingen (as it always did) to ask if it could recommend a replacement. Somewhat surprisingly, the senior staff recommended Johannes Kepler, for reasons that are not quite clear. Perhaps they considered his spirited defence of the Copernican doctrine in a public debate to be not the kind of activity for a priest-to-be, who should be more concerned with the promotion of religious principles than with abstruse, astronomical theories; or, more generously, perhaps they had recognized his exceptional natural gift for mathematics, even though his training in that subject was far below that expected of a teacher in it. Whatever the reason, the result was to have major effect on Kepler's career, and on this story. After considerable hesitation – because of the great change in direction from priest to mathematician, to the great change in institution from the famous University of Tübingen to the relatively unknown provincial school of Graz, and to the great change in location from southwest Germany to southeast Austria on the boundaries of Bohemia – he finally accepted, but somewhat reluctantly, since he made it a condition that he be allowed to resume his studies of divinity at a later date, although this is something he never did. Appointed with the imposing title of Mathematicus of the Province of Styria, he took up his duties in Graz in April 1594.

Kepler's first year in Graz was difficult. He quickly realized that his training in mathematics was inadequate for his task and he spent much of his time studying the subject in order to keep ahead of his pupils, not always successfully. But at the end of that year, while learning and teaching mathematics, he had a sudden inspiration which was to make astronomy his driving passion for the rest of his life. Up to that time, he had been interested in astronomy but not committed to it. While in Tübingen he had read about the heliocentric Universe of Copernicus and had defended it in a public debate, but this was one of his many interests and he did not have the mathematical training needed to enter that field. He never questioned the Copernican system, its simplicity compared to that of Ptolemy being sufficient for him and, like Pythagoras, he was convinced that the Universe would have some form of beauty and order that could be represented by numbers. It was against this background, while studying and teaching geometry in Graz, that he had his great idea. There were five, and only five regular solids (i.e. solids bounded by a number of identical sides), as had been proven rigorously by Euclid. These are, in the order of increasing numbers of sides, the tetrahedron or pyramid (4 equilateral triangles), the cube (6 squares), the octahedron (8 equilateral triangles), the dodecahedron (12 pentagons) and the icosahedron (20 equilateral triangles). One property of such regular solids is that they can be perfectly circumscribed by a sphere which would touch each of their points, and a sphere can also fit perfectly inside them, touching each of their sides. There are six planets with five spaces between them, so Kepler wondered whether it would it was possible that the spheres carrying the planets fitted precisely into the five regular solids, thereby explaining their spacing and the fact that there were only six. Kepler was filled with elation and was convinced that he was right that there was order in the Universe and that it had been discovered by him, a young mathematics teacher in Graz. Of course, we now know that his great idea was a delusion, and that the spacing of the planets has nothing to do with the regular solids and that there are not six but nine planets plus some thousands of asteroids orbiting the Sun. But the delusion was to Kepler a great inspiration, which took him into astronomy for a career which ultimately was to result in one of the major advances in the human understanding of the Universe: the discovery of his famous laws, which were to describe the true motions of the planets.

Kepler now had to subject his new world model to the test of matching the observations, and his first task was to assign his five solids to the five interplanetary spaces. He first thought that they would be ordered in a sequence starting with the one with the fewest sides (the pyramid with four) and ending with that with the largest number of sides (the icosahedron with twenty), proceeding from the outermost planet (Saturn) to the innermost (Mercury), or the other way around. Neither of these fitted and after some time (there are 120 different combinations in the order of the five solids), he came up with the best fit, which was to place the cube within the orbit of Saturn circumscribing that of Jupiter, to place the pyramid within the orbit of Jupiter circumscribing that of Mars, the dodecahedron between Mars and Earth, the icosahedron between Earth and Venus, and the octahedron between Venus and Mercury. This apparently random order (which, in sequence of the number of sides, was 3, 5, 4, 1, 2, going from Mercury to Saturn) was not addressed satisfactorily by Kepler in the book he wrote on his findings, which had a long title which was to be one of his characteristics in writing books:

The mysterium cosmographicum
A Forerunner to Cosmographical Treatises, containing the Cosmic Mystery of the admirable proportions between the Heavenly Orbits and the true and proper reasons for their Numbers, Magnitudes and Periodic Motions
by
Johannes Kepler, Mathematicus of the Illustrious Estates of Styria

Kepler had had a religious training and he intended, in the preface, to write an argument that the Copernican doctrine did not contradict Holy Scripture; but in publishing via the University of Tübingen, he needed the approval of the Dean of the Theological Faculty. The Dean strongly opposed the inclusion of any theological discussion and, in the same spirit as the famous Osiander preface to The Revolutions, recommended that the heliocentric system should be treated as a completely hypothetical proposal that presented a formal mathematical problem. Kepler did leave out the theology but, contrary to the Dean's advice and to the spirit of

Osiander, in the first chapter of the first volume, he made an unambiguous commitment to and belief in the Copernican heliocentric doctrine. He was the first intellectual of note to so do publicly, but there was not the religious backlash that was to occur some 35 years later when Galileo argued in favour of Copernicus, albeit in a totally different manner, in his book *Dialogue on the Great Systems of the World.* Kepler went on to argue in favour of his world model, giving reasons, which were often mystical and always unfounded for the particular order of the regular solids in the orbits of the planets. Indeed, he sought for and invented reasons in favour of his hypothesis, quite contrary to the modern scientific method, of which he was to be one of the forerunners, and his great contemporary, Galileo, the founder. If Kepler's involvement in astronomy had ended there, and if he had remained in that mode of thought for the rest of his life, he would not have a place in this story, but the next stage saw a marked change in his attitude in which his innate sense of scientific rigour overcame his wishful and mystical thinking. He subjected his proposal to the critical test of matching the observations with the criterion that his model should be able to predict the positions of the planets at any one time. The test failed; there was reasonably good agreement for Venus, Earth, Mars and Saturn, but not for Mercury or Jupiter. Kepler did not abandon his concept, which he was still convinced was the right one, and therefore concluded that it must be the observations that were in error; after all, the only ones available to him were those of Hipparchus, made centuries before. But he was well aware of the superb observational programme that had been conducted by the famous Tycho Brahe and, when his book appeared in 1597, he sent a copy to Tycho with a letter of explanation and a request for access to Tycho's data in order to test his model more fully. This started a correspondence, which culminated two years later with Tycho's invitation to Kepler to work with him in the new observatory being built at Benatek.

Kepler's decision to move from Graz and join Tycho was not such a difficult one to take. Apart from the urge driving him to gain access to Tycho's data, he had some major problems in Graz. The Archduke Ferdinand of Hapsburg had started a campaign to cleanse the province of Lutherian beliefs and had closed down the Protestant school in Graz, leaving Kepler without salary and job. Two years earlier he had married the daughter of a rich mill-owner, largely as a result of approaches made by colleagues acting as intermediaries on his behalf and his struggling to cope with the opposition of her father, who considered Kepler to be of too lowly a status and too inadequately paid. It was not a happy marriage; the first two children died in childhood and his wife was so protective of her dowry that she refused her husband any access, even when he had no salary. When Kepler joined Tycho in Benatek, he left his wife in Graz with the intention, which he later fulfilled, of setting up the conditions that would allow her to join him there.

Tycho and Kepler working together should have made the perfect team. The greatest observational astronomer of the age was joined by someone who was to prove to be the greatest mathematical analyst of the age. But although their skills were perfectly complementary, their personalities were not; the association was far from harmonious, possibly because of their very different natures and backgrounds. One was an aristocrat and the other of a lowly family; one was accustomed to wealth, the other was born in poverty and had to struggle financially throughout his career (even in Prague, Kepler's salary was less than 7 per cent that of Tycho); the achievements of one were behind him, the other's were yet to come; one had a gargantuan appetite and was a hard drinker, the other's eating and drinking habits were frugal; and Tycho was well built and robust, Kepler frail and unhealthy. But both had strong beliefs and personalities, one extrovert and the other introvert, and their relations became stormy, with Kepler feeling that he was treated as an inferior, as he was in some instances, for example by being assigned a clearly lowly position at the dinner table. Kepler had joined Tycho with the prime objective of gaining access to his observations (which Tycho had not published), but soon discovered that Tycho regarded them as his own private treasure, which he would release only in limited quantities at his whim, sometimes over dinner. The two were so opposed in nature that only their professional interests kept them together. Kepler wanted, even thirsted after, Tycho's data, but Tycho believed that his observations, representing his life's work, held the secret of the Universe and that Kepler, only Kepler, had the ability to extract it (in this he was right). But what, in so doing, if Kepler and not he, the great Tycho, were to be accorded the historical credit? To an extent, he was justified in this fear because the laws that Kepler was to extract from the observations would bear his name and not Tycho's.

The situation was not conducive to scientific progress. Kepler wanted to be able to pick his own problem and to tackle it his way with unimpeded access to the observations. Tycho wanted him to do what he, Tycho, wanted. At that time, it was known that a major difficulty lay in explaining the motions of the planet Mars, a problem that had defied the efforts of Tycho himself and a senior assistant. The challenge attracted Kepler, who volunteered to tackle it and, although Tycho finally agreed, the data he reluctantly released were inadequate for the purpose. Other disputes arose, particularly concerning Kepler's remuneration and the need to set up accommodation sufficient for him to bring his wife from Graz. After five argumentative months, Kepler left for Graz, not sure if he would return to Benatek or try to re-establish himself in Graz. But he found that conditions there had worsened, the Archduke having stepped up his campaign against Lutherism and issued a decree that gave all Protestants in the province the choice between returning to the Roman Catholic faith with a public declaration, or exile. Like all Protestants, he was called before an ecclesiastical commission, but he declined to change his faith and chose exile. Kepler wrote to Tycho to explain his situation and Tycho responded generously by inviting Kepler to rejoin him as his senior assistant, a permanent position approved by the emperor. But there was a slight sting in the tail; he was to be a true assistant and to carry out only those tasks allotted by Tycho and not those he wished to carry out himself. Kepler had no choice but to accept. With his wife he left for Prague, where the Emperor had asked Tycho to be based, and they arrived there in October 1599, virtually penniless because his wife's dowry was needed to fund the move. The exile, Tycho, was now joined by another exile, Kepler.

Kepler was faithful to his agreement and carried out tasks only as commissioned by Tycho, but none of these took him near his passionate interest, the problem of Mars, and he often found his work boring. Nevertheless, there was now an equilibrium in their relationship, but this was to be ended irrevocably by a tragic development. In October 1601, only 24 months after Kepler's return, Tycho was a guest at a banquet given by a local baron. He was completely at home on such occasions and, as usual, ate and drank prodigiously. He was also fully aware of the etiquette that required that guests did not leave the table except when the host did and, on this occasion, he was finding difficulty in meeting it. Unfortunately for Tycho, the baron did not need the comfort break that would have allowed him to follow suit and he had to wait until the end of the banquet for relief, when he found great difficulty in urinating. His carriage took him home where he went to bed in some pain (presumably caused by a urinary counter-flow), and he died 11 days later. In describing the incident later, Kepler remarked that "Tycho had held back his water beyond the bounds of courtesy", a sentiment with which I am sure you will agree. Tycho was buried in Prague and, two days after his funeral in early November 1601, Kepler was appointed his successor as Imperial Mathematicus. He finally had his unfettered hands on Tycho's treasure of astronomical data and he was to exploit these superbly by discovering his three famous laws which described, for the very first time, the motions of the planets – a major advance in the cosmology of time.

Kepler immediately embarked on the study of Mars that Tycho had impeded. It should be remembered that he was still committed to the idea that the planets were spaced according to the five regular solids, so his first task was to determine their orbits, starting with Mars, which, on Kepler's scheme, should fit inside a pyramid and circumscribe a dodecahedron. But this first task was to become a major challenge and his idea of the regular solids began to sink further and further into the background.

He was diverted from this work temporarily in 1604 when another new star appeared, which he was to study, write a book about, gain immediate fame from and have his name attached to. This is a story very similar to that of Tycho's supernova, which had occurred only 32 years earlier. It is quite remarkable that the only two supernovae that have occurred in the Milky Way during the past thousand years happened in the early stages of their careers, enabling both of them to establish themselves as astronomers. But their great achievements were to lie elsewhere in astronomy and were attributable to the combination of the precise observations of one and the penetrating analysis of the other.

Kepler returned to the problem of Mars, which he confidently thought would take him only eight days; after all, he was only repeating the calculations of Copernicus, but with the more accurate data of Tycho. In the event, it took Kepler eight years and nearly a thousand pages of closely written calculations before he cracked the problem and discovered his first two laws of planetary motion (the third was to wait another nine years). The delay was caused by not only the tedious and laborious nature of the extensive numerical calculations that were needed, but also by the psychological problem of shedding one of the sacrosanct dogmas of the past, namely, that the motions of the heavenly bodies must be in circles at constant velocity. Pythagoras, Aristotle, Ptolemy, the Islamic astronomers, and even Copernicus, had accepted this as a basic premise. Kepler also started from this basis, but even though he adopted some of the little fudges of previous calculations, such as displacing the Sun from the precise centre of the planetary orbits (with which he was very unhappy), he could not find an orbit for Mars that matched the precision of Tycho's data (about one minute of arc). He could find orbits that matched the precision of the observations available to Copernicus (about one degree) but not the precision of Tycho. At this point, Kepler acted with the rigour of modern science (an innovation for that time) and realized that he would have to abandon the long-established dogma of circular motion at constant velocity. He abandoned the latter first and improved his fit to the data, but only when he also abandoned the former did he get what he was looking for, an orbit and velocity that matched Tycho's data perfectly.

It was not easy. If the orbit of Mars was not a simple circle, but what of the many possible shapes could it be? If the velocity was not constant, what form of variation could it take? Kepler's tools were limited to arithmetic and geometry and he had to use his insight to test out different kinds of orbit and different types of velocity variation, each requiring elaborate calculations. After the greatest numerical analysis ever undertaken, he published his findings in 1609 in his magnum opus which, like *The Mysterium*, bears a rather lengthy title:

A New Astronomy
Based on Causation
or
A Physics of the Sky
derived from Investigations of the
Motions of the Star Mars
Founded on Observations of
The Noble Tycho Brahe

The acknowledgement of his debt to Tycho, displayed prominently in the title, was repeated by his statement in the text that the study had started under "Tycho's supreme command". Of course, he knew that his great achievement would not have been possible without Tycho's superb observations, and his acknowledgement was a genuine one that revealed a touch of generosity in Kepler's character not present in all scientists or philosophers.

The *New Astronomy* gave the arguments and proof of the discovery of his first two laws of planetary motion. The first law stated that the orbit of Mars and, by implication, all of the other planets, was an ellipse, with the Sun located at one focus. An ellipse is a shape obtained by slicing a cone with a plane at an angle to its base. It is a symmetrical and well defined figure with two foci and has the property that the length of a string from one focus to any point on the ellipse, and then to the other focus, is constant. Also, the normal to any point on the ellipse bisects the lines to it from each focus; this means that a point of light at one focus will produce a perfect image at the other, because the law of reflection is obeyed at each point. The earlier attempts to fit the data by displacing the Earth (by Ptolemy) or the Sun (by Copernicus) were explained because each focus is displaced from the geometric centre, which lies equidistant from the two. His second law states that the velocity of a planet varies in such a way that the line joining it to the Sun sweeps out equal areas in equal time. Hence, when the planet is nearest to the Sun (its perihelion) it moves most quickly and when it is at its farthest distance (its aphelion) it moves most slowly.

Although it took Kepler another nine years to discover and publish his third law of planetary motions, it will be dealt with here for completeness. This law concerned the relation between the periods of the planets and their distances from the Sun. The former were known precisely because of the large number of revolutions that had been observed since ancient times, but their distances were extremely poorly known, so much so that there was no basis for determining how the periods varied with distance, a question of fundamental importance that he had asked himself in *The Mysterium*. In order to answer the question, he would have to tackle one of the central problems in astronomy – the determination of the distances of the heavenly bodies – an issue that had not advanced one iota since the work of Aristarchus in the third century BC. Kepler was to develop methods to determine the distances of the planets, the first to do so, once again exploiting Tycho's observations and once again demonstrating brilliant analytical skills; he was to be rewarded by the discovery of his third law.

Kepler's methods were geometrical and they consisted of defining a triangle which had the Earth and the particular planet at two corners and another appropriate body at the third corner. If the length of one of the sides (the baseline) is known and two of the angles measured, then the Earth–planet distance could be calculated from simple geometry; he would then transform it to a Sun–planet distance. These techniques are basically the same concept as in surveying: a baseline is set out and the angles towards the object being surveyed are measured from each end of the baseline, allowing its distance to be calculated; the longer the baseline, the more accurate the method. In determining the distance to the Moon, described earlier, Aristarchus essentially used the Earth's diameter as a baseline and, since this had been measured accurately by Erastothenes, and since the Moon is only about 30 Earth diameters away, the result was a good one. However, when Aristarchus attempted the measurement of the distance to the Sun, this was about 400 times longer than his adopted baseline, the Earth–Moon distance. This, together with the fact that the angle he had to determine (when the Moon was half full) could not be measured as precisely as the position of a star or planet, resulted in a grossly inaccurate estimate. Kepler, with his usual insight, was to adopt a much longer baseline, the distance from the Earth to the Sun, a distance we now call an astronomical unit (AU). Since the most distant planet then known (Saturn) was only ten times farther from the Sun than the Earth, the baseline was very adequate for the purpose, but, as already mentioned, the Earth–Sun distance was extremely poorly known and was to remain so until the second half of the seventeenth century, when a new geometrical method of measuring the Earth–Sun distance was developed and applied to give the first reasonable estimate; as will be recounted later, this has been improved with time until today we know it precisely. Kepler's measurements of distance were therefore in terms of astronomical units, the mean Earth–Sun distance, but they were accurate in the relative sense and this was sufficient to establish his third law.

Kepler developed different methods for measuring the distance to the inner planets and the outer planets. These were based on the heliocentric model and were applied by using Tycho's data. For the inner planets, Mercury and Venus, he selected from Tycho's observations those that were made when the planets were at their maximum elongation from the

Sun, because he knew that, in such a situation, the angle between the Earth–planet line and the planet–Sun line was 90°. Since he could also measure the elongation – the angle between the Sun and planet – he had defined his triangle and could determine the Sun–planet distance in terms of the Earth–Sun distance.

For the outer planets, Kepler had to develop a different solution. He exploited the fact that the periods of the planets (the times they take to orbit the Sun) were known precisely, because they had been observed over many thousands of years. The principle is the same for each of the outer planets then known (Mars, Jupiter and Saturn) and here Mars will be used to demonstrate the method. From Tycho's data, Kepler extracted those observations which were exactly one Martian year (about 1.5 Earth years) apart. He therefore knew that Mars was at the same position in its orbit, but that the Earth was in two different positions. Since, by geometry, he could express the separation of the two Earth positions in terms of its distance from the Sun, he had established a baseline by which the angles to Mars allowed its distance to be determined. He exhaustively used all of Tycho's observations relevant to the problem and derived, for the first time with any accuracy, the distances of the planets from the Sun in terms of the distance of the Earth from the Sun, the astronomical unit.

With his distance measurements, Kepler quickly established his third law of planetary motion, which gives the relationship between the period of revolution of each planet around the Sun and its distance from it. It states that the square of the period is proportional to the cube of the distance. This means that the more distant planets take longer to orbit the Sun, not only because they have farther to go, but also because they move more slowly. Kepler completed this great work in 1618 and published it in his book *Harmonice Mundi* (*The Harmony of the World*) which appeared in 1619.

In summary, Kepler's three laws of planetary motion are:
- Planetary orbits are ellipses, with the Sun located at one focus.
- A planetary velocity varies in its orbit such that a line from it to the Sun sweeps out equal areas in equal time.
- The square of the period of a planet is proportional to the cube of its distance from the Sun.

These three laws are the great achievement of Johannes Kepler. He had discovered them buried in the extensive set of observations made by Tycho Brahe and had extracted them with the most brilliant series of mathematical analyses that had ever been conducted at that time. His interest in astronomy had been inspired by the conviction, also held by Pythagoras, that there must be some kind of order in the motions of the heavenly bodies that could be expressed mathematically. As a young mathematics teacher in Graz, he had made his first excited guess that the orbits of the planets were such as to fit into the regular solids of geometry, of which there are only five. His work was driven by the wish to prove this hypothesis to be correct, but it did not deviate him from his search for the truth, which he knew only the observations could reveal and, as the distinguished Imperial Mathematicus of Prague, he discovered his three laws. With these he proved that the general belief he had held, with Pythagoras, was correct: there was a harmony in the sky, there was an order in the Universe. The order was totally different from his early hypothesis of the five regular solids, and he did not understand its reason or purpose, but order it undoubtedly was.

In contrast to his professional achievements, Kepler's domestic life could not be described as happy or prosperous; indeed it was subject to difficulties and tragedies, particularly in the last 20 years of his life. These started in 1611, which was a great year for English literature with the appearance of the authorized version of the Bible, commissioned by James I of England, one of the finest examples of the English language ever written, even rivalling Shakespeare, who produced *The Winters Tale* and *The Tempest* in the same year. But, for Kepler, it was a deeply tragic year; the plague was rife and it struck down his three surviving children (the first two died in early childhood) finally claiming one of them, his favourite; his wife caught a dangerous fever and, after a period of mental imbalance, died at the age of 38. In a struggle for the throne, an armed insurrection broke out, which led to the abdication of Rudolph and his replacement by his brother, Mattias, who was elected King of Bohemia. Without his patron, Kepler moved back to Austria where he once again took up a lowly position as teacher of mathematics in a provincial town, Linz (where Adolph Hitler was later to attend school), but the new king allowed him to retain his title of Imperial Mathematicus. In 1618, the long smouldering struggle within Christianity between the breakaway Protestant faiths and the established Roman Catholic Church, between the Reformation and Counter-reformation, broke out into armed conflict with the bloody Thirty Years War. At that time, Europe (and the New World) was also subjected to one of those bouts of madness which too often beset the human race. Witchcraft was at its zenith and almost claimed Kepler's mother, who was formally accused of that terrible crime and whom he defended during her 14 months of imprisonment before her trial in 1621; she was acquitted, largely because of Kepler's efforts, but died six months later. Kepler was usually in financial difficulties during the whole of his life and often wrote articles or accepted commissions in order to supplement his income. But he was always loath to press for payment and, since the costs of his mother's trial had increased his financial problems, he embarked on a long journey in 1630 to collect a longstanding but substantial debt, but took ill and died en route.

There is one more event to relate in the life of Johannes Kepler, which occurred in March 1610, one year after his *New Astronomy* was published. A close friend and admirer, who was also a Privy Counsellor to his Imperial Majesty, arrived at Kepler's residence in a state of great excitement. News had arrived at Court that an Italian mathematician, Galileo Galilei, had built a new instrument, based on a Dutch invention and called a telescope, which he had pointed at the sky and thereby discovered mountains and valleys on the Moon, countless numbers of hitherto unseen stars and, wonder of wonders to Kepler, four new moons revolving about the planet Jupiter – a miniature Copernican system. Of wider moment, it heralded a decisive landmark in the development of astronomy, with the introduction of the telescope; an immensely powerful tool was now available to probe the secrets of the Universe.

Galileo was born in 1564, the same year as Shakespeare, and his contributions to science were to match those made by Shakespeare to English literature. He was to become regarded as the father of modern science by establishing the scientific method, by which all hypotheses, all theories and all claims should be subjected to experimental test and verification. Further, the explanation of existing knowledge by any theory was not in itself fully sufficient, because it needed to predict new facts and be continually subject to the test of any new developments. This principle, which seems to be supremely self-evident, was nevertheless regarded with hostility by the establishment of the time. Indeed, although it is a philosophy from which all areas of human activity would benefit, it is far from being universally adopted outside science, even today. In politics, the same set of facts can be used selectively by opposing parties to reach diametrically opposite conclusions, usually with great skill. But, of course, scientists are not more rational than other groups but are subject to the same human emotions, neither are they without the normal human frailties such as prejudice, self-interest, and so on; indeed, in my experience these are more abundant than in other professions. However, scientists do subject their propositions and theories to the discipline of independent, rigorous tests. This scientific method, heralded by Kepler, defined and demonstrated by Galileo, and to be triumphantly exploited by Newton, was to lay the basis for the understanding of nature and the great human technological developments that led to the achievements and comforts of modern society. One of the problems that resulted from adopting the discipline of the scientific method is that science and technology advanced at a rate greater than in social and political areas, and this resulted in scientific and technological advances of clear benefit to the human race, being opposed because the political and social capabilities could not match the technological advances. An illustrative example is afforded by the Luddites, who destroyed newly introduced textile machinery in early nineteenth-century England in order to protect the jobs of hand-loom operators. Examples continue to arise today. Its philosophy was in complete antipathy to that of Aristotle, and it was also totally foreign to the established dogmatic way of thinking in Galileo's time which looked for truth and certainty in ancient writings, both scriptural and philosophical, and opposed any questioning of them. It is therefore not surprising, indeed some might say that it was inevitable, that a collision should occur between Galileo and the establishment in the form of the Church. But that collision was not entirely attributable to the diametrically opposed philosophies but also, at least partially, to the personality of Galileo. It is interesting and somewhat appropriate to consider the comment of another astronomer, made several centuries before Galileo lived, and expressed in English by a poet two centuries after he died:

But leave the wise to wrangle, and with me
The Quarrel of the Universe let be.

The astronomer was Omar Khayyam, who was mentioned earlier, and the poet was Edward Fitzgerald. But it is certain that, had the advice been known to Galileo, he would either have ignored it or written a sharp, sarcastic response to demolish it.

Galileo started his education at a Jesuit school near Florence. His father wanted him to become a merchant, but, when his intellectual ability became apparent, he decided to send him to study medicine at the local University of Pisa, which Galileo entered at the age of 17. His father was not rich and he had to support five children, so, with university fees being high, he sought a scholarship for Galileo which, remarkably, he failed to win. There is a similarity here with Kepler, who could not have gone to university at all without the scholarship he won; the same financial stringency had now overtaken Galileo and, since he did not win a scholarship, he had to leave university without a degree. While at Pisa, the first signs of his scientific curiosity had emerged, stimulated by his observations of the chandeliers swinging slightly in the air of the cathedral. From these, he conducted some of the first true scientific experiments and established that a pendulum of some given length would oscillate with exactly the same period of time whatever its amplitude and whatever the weight or nature of the suspended body. He had established the principle of the first mechanical chronometer and he went on to build a "pulsilogium" for measuring human pulse rates, a reflection of his interest in medicine, which arose out of one year of university study. He went on to demonstrate his inventive ingenuity by developing other instruments, such as a hydrostatic balance which received such attention that he was appointed a lecturer in mathematics at Pisa, without a degree in anything and four years after being refused a scholarship by that university for entry as a student. His continuing achievements led to his being offered, three years later, an appointment as Professor of Mathematics at the renowned University of Padua. He accepted, at the age of 28, and was to spend 18 of the most fruitful years of his life there. He was to establish the foundations of the science of dynamics on an experimental basis and to make a major breakthrough in astronomy.

Galileo quickly discarded Aristotle's concepts of motion and gravity. He showed that a body thrown into the air with transverse motion would not fall vertically downwards on return, but would retain that transverse motion; hence, Aristotle's horseman at full gallop would find that a weight that he hurls upwards would not fall behind him as it would if it returned precisely vertically, but that it would retain its lateral velocity and fall back on to him, except for any small difference caused by wind resistance. You will remember that this was an experiment that Aristotle did not have carried out since, being a foregone conclusion, it was quite unnecessary. Of the other experiments that Aristotle did not conduct, using the same argument, there was one of supreme importance to science and astronomy; the test of the statement that heavy objects fall faster than light objects, such that a body twice as heavy as another body would fall twice as fast. Following his philosophy, Galileo did conduct this experiment in the laboratory by using balls on inclined planes to slow down the motion and thereby allow more accurate measurement; the result was that all balls fell equally, whatever their weight or size. More theatrically, he dropped two very different weights from the Leaning Tower of Pisa (it does not matter whether this story is apocryphal) and both struck the ground at precisely the same time. Galileo had established the basic and unique property of gravity – all objects, whatever their size, whatever their mass, whatever they are made of, fall identically in a gravitational field. This property has been tested ever since and, in modern times, has been shown to be correct to an accuracy of one part in 100000 million. Of course, in the environment of our atmosphere, different degrees of air resistance can cause a difference in rates of fall, of which the most extreme example is a feather which floats rather than falls to the ground. But in the absence of air resistance, when only the force of gravity is present, a feather will fall as fast as a heavy object. This was to be demonstrated by Isaac Newton, also theatrically, when he evacuated a tall glass tube which had a golden guinea and a feather supported by a hinged ledge at the top. On dropping the ledge, the guinea and feather fell identically and struck the bottom at the same time. This property, which is unique to gravity among all of nature's forces, was to be used first by Newton and then by

Einstein in developing their respective theories of universal gravitation and general relativity.

Galileo's impact on astronomy was immense and dramatic. While Professor of Mathematics in Padua, he heard of a recent Flemish invention of an optical instrument which had the property of increasing the power of the human eye, thereby allowing distant objects to be magnified and appear to be much nearer. It was called a telescope and, quickly working out the principle, he designed and built one himself. It had a magnification of ten and he demonstrated its power from the Tower of San Marco in Venice, from where he could see ships 32 kilometres or two to three hours sailing time away. This had great economic value because it gave sufficient notice for the unloading of cargo to start immediately on docking, and it also had great military value because a hostile fleet could be detected far enough in advance for full preparations to be made for defence. This brought him considerable fame and some fortune, because a grateful Venetian Senate immediately doubled his salary and also offered to provide full funds to construct an improved version. Galileo then built a much improved, more powerful telescope, which gave a magnification of 33 and, with an aperture of nearly five centimetres, effectively increased the power of the human eye by a factor of a hundred. But he did not use it for the commercial and military reasons of his paymasters; instead, he pointed it at the sky and made the first telescopic observations of celestial objects, thereby establishing a milestone in astronomy. A new powerful instrument had been introduced which was to revolutionize our knowledge of the Universe.

Galileo's great astronomical discoveries were made by exploiting his telescope with a keen eye and a brilliant mind. He described the wonders revealed by his telescopic observations in a book published in 1610 under the title *Siderius Nuncias (The Starry Messenger)*. It caused a sensation in the learned world and gave him fame far beyond that which he had achieved in the merchant world of Venice. The book described and gave drawings of mountains, craters and sea-like areas (maria) on the Moon; although the stars still appeared as points, the planets were seen as disks; Saturn appeared spherical like Mars but had considerable extensions beyond its disk and Galileo's observation constituted the first detection of that planet's rings. Of considerable cosmological importance were his observations of Venus and Jupiter. Over time, Venus showed phases like the Moon, varying from a thin crescent to being almost full; this clearly demonstrated that it was visible by reflected sunlight and that it was revolving about the Sun and not the Earth. In the case of Jupiter, Galileo detected four new planets or, more accurately, moons, which lay in the same plane and, over a period of time, revolved around the planet – a Copernican system in miniature. These observations demolished the Aristotelian view that all heavenly bodies moved in circles with the Earth at the centre. Although Galileo's observations were consistent with a Copernican model, they were also consistent with the model of Tycho Brahe, which allowed the planets to revolve around the Sun and, by implication, the Jovian moons to revolve around Jupiter, but the Sun, Moon and stars still revolved around the Earth. This therefore retained the central Aristotelian thesis of a fixed, stationary Earth as the centre of the Universe, a thesis that was to be the subject of bitter dispute and argument.

Galileo's observations of the stars also had a cosmological implication. Wherever he pointed his telescope, he could see far more stars than could be seen with the naked eye, particularly in the Milky Way, where they appeared in countless numbers. The Aristotelian view was that the stars were fixed in a sphere which marked the boundary of the Universe and whose centre was located at the Earth. This required the countless number of additional stars to be intrinsically fainter if they were located in that sphere but also raised the more plausible explanation that they were simply more distant. This was the first inroad into the thin, shell-like, spherical Universe of Aristotle.

Galileo's fame now gave him easy entry to the Courts of Italy and to the Vatican, which he visited several times. Always a strong supporter of the Copernican theory, his own observations had made him even more so, and in discussions and letters he promulgated the heliocentric view. This caused opposition among many intellectuals in the Church, and the debate increased in intensity until 1632 when Galileo published his book *Dialogue on the Great World Systems* in which the pro-Church view was put into the mouth of a character called Simplicius. The pro-Copernican argument was masterful, the polemic was brilliant, but a little cruel, and the words and characteristics of Simplicius were argued by some of Galileo's enemies to bear some resemblance to the Pope. It is probably the latter that triggered off one of the unfortunate events of science history, the trial of Galileo. The book was banned by order of the Pope, and Galileo was summoned to appear before the Holy Inquisition. This he did on 12 April 1633 at the age of nearly 70, when he was to be charged with heresy on the grounds that he taught the Copernican doctrine, whereas

The proposition that the Sun is the centre of the world and does not move from its place is absurd and false philosophically and formally heretical, because it is expressly contrary to the Holy Scripture.

The proposition that the Earth is not the centre of the world and immovable but that it moves, and also with a diurnal motion, is equally absurd and false philosophically and theologically considered at least erroneous in faith.

The trial of Galileo should be viewed in the context of the age in which it occurred. In 1633 the religiously inspired Thirty Years War was at its peak and had spread through most of continental Europe; Heidelberg fell to the Protestants, who also advanced into Bavaria; on the other side of the world, the Colony of Connecticut was founded, an event not totally unconnected to the religious divisions within Christianity at that time. But the trial of Galileo was to be seen, then and since, in a much wider context than the reformation, counter-reformation or the specific issue of the Copernican doctrine. It was to be seen as a conflict between Church and science, a conflict that was perhaps inevitable because of the basically different approaches to understanding the natural world, a conflict that had been declared unambiguously by the Christian theologian Tertullian some 14 centuries earlier, and in a tone that was remarkably prophetic and apposite to the trial of Galileo:

This is the substance of secular wisdom that it rashly undertakes to explain the nature and dispensation of God . . . Heretics and philosophers deal with similar material, and their arguments are largely the same.

Although his personality and bearing ill suited him for the part, Galileo was destined to be the hero, the defender of the theory of someone else who had been dead some 90 years, the individual against the establishment, the defender of experimental facts against dogma and of the belief that laws govern nature against the belief that everything is God's whim as defined by Holy Scripture and, therefore, cannot be challenged. But he was bound to lose. The trial took place during the middle of the Thirty Years War, when the split in Christianity had flared up into bloody conflict and, although Galileo had no part in that split or conflict, the climate of the time was ruthlessly against any challenge to the Church, either real or imagined. In this case, it was undoubtedly imagined; Galileo was a loyal member of the Roman Catholic Church; he did not dispute the scriptures but argued that the clerical interpretation of them was wrong, a point of view that was finally conceded by the Vatican 360 years later. Neither was Galileo's trial any part of the religious argument between Protestant and Catholic; remember that the preface to Copernicus's *Revolutions*, which was written by a prominent Lutherian theologian, expressed the same formal attitude to the heliocentric doctrine as was held by the Catholic Church at the time of Galileo.

Galileo's trial was conducted by ten judges in the normal manner of the Inquisition at that time and was a violation of all the principles of justice now generally accepted in democratic countries. The accused was not informed of the charges against him until some appropriate time during the trial; he was not allowed to see any proof of evidence brought against him; he was not allowed a defending counsel. The inquisitors addressed him in the third person. How did he come to Rome? Was this his book? Why did he write it? Why did he think he had been summoned before the Holy Inquisition?

Galileo answered the last question by saying he believed it was because of teaching the Copernican system, and then advanced a lengthy argument in its favour. He was fully prepared to deal with the counter-arguments (as he had in his book), but then came the unexpected questions: Was he in Rome in 1616? Did he meet with Cardinal Bellarmine? What did he learn from His Eminence? Galileo responded that the Cardinal had explained the formal position of the Church in respect of Copernicus, which was that to hold the Copernican theory to be a proven fact was contrary to Holy Scripture, but that it could be used as a hypothesis. This was indeed the official position of the Church, and Galileo had a certificate to that effect. The question seemed to rest on whether the Dialogue was presenting the Copernican model as true fact or as a hypothesis for intellectual discussion. The issue was a fine one but it seemed at that point to be a matter of life and death to Galileo.

It was at this stage that the Inquisition revealed its hand and based its prosecution on one issue. During 1616, Galileo had been instructed by the Holy Office "to relinquish altogether the Copernican theory, nor henceforth to hold, teach or defend it *in any way whatsoever*, orally or in writing" [author's italics]. There was no fine point in this unambiguous statement. A document to that effect, the Inquisition claimed, had been signed by Cardinal Bellarmine before a notary and witnesses, and also by Galileo. It was not produced (it did not need to be), nor has it been found since. But its implication was clear; if Galileo had been given such an instruction, then his book openly and publicly disregarded it, and was therefore a basis for a heresy charge.

Galileo must have now seen the writing on the wall and, as the trial moved into its next phase, chose not to take the ultimate and probably fatal path of the martyr. The instruments of torture were displayed before him and detailed explanations of how they would be applied were delivered, together with the persuasive argument that a full confession would avoid such pain. He was also conscious of the experience of a previous committed Copernican, the Dominican monk Giordano Bruno, who was also very outspoken. For example, the preface to *The Revolutions* says, among other things, that "the doctrines therein were purely hypothetical and not meant to represent reality"; this led Bruno to write that "it could only have been written by one fool for the benefit of other equally ignorant fools". He was tried, found guilty of heresy, and burned at the stake in 1600.

Galileo confessed, totally and abjectly, and recanted his belief in the

Copernican theory. It is often reported that, as he was led away, he defiantly muttered "but it [the Earth] still moves"; this may be apocryphal, but is consistent with the fact that his belief in the Copernican theory was undiminished. At this point, it should be remembered that the inquisition sentenced him, a convicted heretic, to the lightest sentence possible. For that crime, many people had suffered death by burning at a stake, a death as horrific as that suffered by Jesus Christ, in whose name it was being carried out. Galileo was to be confined for life in his villa in Arcetri, Florence where he was well provided for, but was forbidden to publish anything or to talk to Protestants. This latter constraint again demonstrated the confusion of the Roman Catholic Church between its argument with a scientific approach to understanding the natural world, rather than a scriptural one, and its conflict with the Protestant reformation; these were two totally different issues.

Millions of words have been written about the trial of Galileo, so I apologize for adding a few more by way of comment, admittedly with a few centuries of hindsight. First, from a purely scientific point of view, the information available at the time did not form an unambiguous support for the Copernican doctrine. Galileo had clearly demonstrated that Venus orbited the Sun and not the Earth and that his newly discovered moons orbited Jupiter and not the Earth. These observations made severe inroads into Aristotle's cosmology, but were consistent with Tycho Brahe's modification of that cosmology, which allowed the planets to orbit the Sun but retained a fixed, stationary Earth as the centre of the Universe. Whether this was true, or whether the Earth also moved around the Sun and with a diurnal motion, was the central issue in the argument. As already said before and repeated now, the crucial test lay in the detection of stellar parallax, because any motion of the Earth around the Sun would cause an apparent wobble in the position of the stars. This had not been detected by either of the great observational astronomers, Hipparchus and Tycho Brahe, and led both to maintain a geocentric Universe. But there was one other argument in favour of Copernicus which, although indirect and aesthetic, was nevertheless extremely strong. This had been provided by Johannes Kepler in the form of his three laws, and the argument that develops from them is: if we assume that the Sun (not the Earth) is the centre of the Universe, then we find the following to hold:

- The planets move in perfect ellipses with the Sun (not the Earth) at one focus.
- The planets move at varying speeds, such that the line to the Sun (not the Earth) traces out equal areas in equal time.
- The periods of the planets depend on their distances from the Sun (not from the Earth) in the form that the square of their periods varies with the cube of their distances.

These clearly place the Sun and not the Earth in a privileged position, and it was the order and the simplicity of the system that was convincing to the scientific mind; it is a little surprising that it was not equally convincing to the clerical mind, since such order could easily have been interpreted as a sign of the hand of God. This leads naturally to viewing the trial of Galileo from a purely clerical point of view and, since the author is ill equipped to do so, he defers to the Catholic Church itself, which has deeply and extensively considered the issue, again retrospectively. At the time of his trial, Galileo's writings were "indexed", that is they were forbidden to all Roman Catholics but, recognizing the build-up of scientific arguments in favour of Copernicus, Pope Benedict XIV, more than a century later in 1757, granted an imprimatur on all of Galileo's works, removing them from the forbidden index. In 1981, Pope John Paul II set up a study commission of the Pontifical Academy of Sciences, with the task of thoroughly reviewing the Galileo trial and its conclusion. This reported in 1992 (after 11 years of deliberation) to give the considered view that Galileo's sentence was an error of judgement based on a dogmatic misinterpretation of Holy Scripture and that the mistakes of that time should be frankly admitted by the Roman Catholic Church. This was done by Pope John Paul in an address in which he stressed the complementarity of the two different ways of gaining knowledge, one by reason and observation, the other by faith. Galileo had been rehabilitated, and his verdict of guilty to the charge of heresy had been reversed, 360 years after it was delivered.

After his trial, Galileo spent the remaining years of his life in considerable comfort in his villa near Florence. He extended his studies of the motion of bodies that he had conducted in Pisa and compiled these in a masterful book, *Dialogues concerning two new Sciences*, which essentially became the basis of the science of dynamics. Against the instruction from the Church, it was smuggled out and published in Leiden before his death in 1642. On Christmas Day of that same year, the person who was later to say that he stood on the shoulders of giants (meaning Galileo and Kepler) "in order to see a little farther than anyone else" was born in the village of Woolsthorpe, Lincolnshire to the Newton family; they christened the child Isaac.

Isaac Newton was born a few months after his father died and, after his mother remarried, he was brought up by his grandmother. Never an intimate person, he had no close friends and, like Galileo, never married (but unlike Galileo, he did not father several children). Little is known of him as a schoolboy or undergraduate and he was a very solitary individual who did not seem to need, like most scientists, the intellectual stimulation that comes from interacting with other sharp minds. His first great achievements were made in his mid-twenties when working in the total intellectual isolation of his home in Woolsthorpe during 1666–7, when his university, Cambridge, was closed because of the plague. That inclination to work on his own continued throughout his life but, nevertheless, Newton was to become the greatest scientist of his age and arguably of all time. He made immense contributions to optics by inventing and building the first reflecting telescope, and by demonstrating that sunlight could be split into a spectrum of different colours – two achievements of immense importance to the development of observational astronomy, which will be described later. But the first and main achievement to describe is how he took Galileo's concepts of the motion of terrestrial bodies, created a new mathematics, and developed a cosmological theory based on his universal law of gravitation, which fully and elegantly explained the motions of the planets as given by Kepler's celestial laws.

A good starting point is the story of the apple, which he observes in his garden (like Galileo's Tower of Pisa experiment, it does not matter whether it is apocryphal). The apple is stationary but when the stem breaks, it falls to the ground; therefore there must be a force (gravity) which propels it. But why should the force be confined to the Earth? What if it extended as far as the Moon? But the Moon did not fall to Earth like the apple; however it was revolving around the Earth, and Newton argued that, if he hurled the apple horizontally as a projectile, the faster it was propelled, the farther the distance before it struck the ground. If he could throw it fast enough, it would reach a speed (ignoring air resistance) where its "falling" matched the shape of the Earth; it would then be in orbit. Newton knew the size of the Earth and he had measured (as had Galileo) the acceleration caused by gravity, so he was able to calculate the period of such an orbit to be 90 minutes. He also knew the distance to the Moon (about 60 Earth radii) and was therefore able to calculate what its period would be if the force of gravity were universal, but diminished as the inverse square of the distance. The answer he obtained for the period of the Moon was 29 days, in agreement with the known value, and this must have induced an intense feeling of elation and excitement. By calculating the period of the Moon using only Earth-bound measurements, he had projected a terrestrial law out into space, the first time this had ever been done, and it represented intellectual courage of the highest order.

Newton carried out these calculations with the same mathematical tools that were available to Kepler – arithmetic and geometry – and realized that these were completely inadequate to determine the result of his law of gravitation on the motion of heavenly bodies. He therefore created a new mathematics to deal with instantaneous motion, which he called "fluxions", but which we now call calculus, one of the most powerful math-

83

ematical tools ever invented. It was also developed, quite independently, by Baron Gottfried von Leibnitz, a contemporary of Newton, but Leibnitz did it as a mathematical challenge, whereas Newton did it to solve his world model.

Newton started his monumental work by setting up three laws of motion, which he derived largely from the terrestrial experiments and arguments of Galileo. The first is the *law of inertia*, which states that, in the absence of external forces, a body will move at uniform velocity in a straight line. (This is completely contradictory to Aristotle, who claimed that the natural state of any body was stationary.) The second is the *law of acceleration*, which states that the acceleration induced in any body is proportional to the force applied to it and inversely proportional its inertial mass; it is the *inertial* mass that is used, because this represents the degree of resistance that a body has to any change in its motion. Finally, the third law is the *law of reaction*, which states that for every action, there is an equal and opposite reaction; a familiar example is a jet engine, whose action propels the aircraft forwards, but whose reaction is expended on the air.

It was from the first law that Newton's ideas of gravity developed, because neither the Moon nor the planets were travelling in straight lines but in very curved orbits (ellipses, as Kepler had discovered). Hence, they must be subject to a force and he proposed that this force was gravity, which operated not only on the ground but which extended universally through space; he had taken the giant step of proposing action at a distance without any material connection. His simple calculation of the Moon's period had told him how the force varied with distance and he constructed his *law of universal gravitation*, which states that an *attractive* force exists between any two bodies which is proportional to the product of their *gravitational* masses and inversely proportional to the square of the distance of their separation. By Newton's third law, this force is the same on both bodies but in opposite directions. If the gravitational mass of one body is doubled, the force is doubled, and if the gravitational mass of both bodies is doubled, the force increases by a factor of four. If the distance between the two bodies is doubled, the force is reduced by a factor of four.

Newton was now able to put his force of gravity into his second law of motion in order to determine the consequences for any body in a gravitational field – the falling apple in his garden, or the Moon in the Earth's gravitational field, or the planets in the Sun's gravitational field. In the case of the falling apple, Newton's second law of motion and his law of gravity gives its acceleration as being proportional to the product of its gravitational mass and the gravitational mass of the Earth, and inversely proportional to its inertial mass and the square of its distance from the

Earth's centre, that is, the Earth's radius. Gravitational mass is the mass that causes the attractive force of gravity; inertial mass is the mass by which a body resists any change in its motion, even in the total absence of a gravitational field. Newton now introduced the special property of gravity that had been demonstrated by Galileo (and confirmed by himself), namely that all bodies, whatever their mass or their nature, fall equally in a gravitational field. In Newton's theory, this meant that the inertial mass in his second law of motion was equal to the gravitational mass in his law of gravity; hence, since the acceleration of the apple is proportional to one and inversely proportional to the other, the mass of the apple cancels out giving an acceleration totally independent of it, in keeping with Galileo. Having measured the acceleration of his apple (or any other body) attributable to gravity, Newton was able to estimate the value of the universal constant of gravitation in his law of gravity. Hence, his laws of motion and gravity were a terrestrial development and he was now armed with the tools to tackle the great celestial problem of the motion of the heavenly bodies.

With his new and powerful tool of mathematics ("fluxions"), Newton was able to transform an acceleration on a body to the resulting velocity of that body and further transform it to its position, all in algebraic terms as a continuous function of time. He was now able to determine the full trajectory of a body in the gravitational field of another body, in which the final solution depended only on the initial condition of the body: stationary in the case of the apple, but with transverse motions in the case of the Moon and planets when they were formed.

Newton then used his calculus to apply his laws of motion and gravity to the specific problem of planetary motions. He treated all planets as being only in the gravitational field of the Sun, an approximation, but a very good one because the Sun comprises all but 1 per cent of the mass of the Solar System. But all bodies gravitationally attract all other bodies, and the small effects of the attraction between planets, which Newton justifiably neglected, were to assume importance in the nineteenth century, as will be described later. The general solution of the orbits of bodies in the gravitational field of the Sun was a conic section, which is the shape caused by slicing a plane through a cone. If the slice is parallel to the base, it is circular, but when it is slightly inclined, it is elliptical. This was the situation for the planets, which were formed by material spiralling in under gravitational contraction and therefore had velocities close to but not exactly at right angles to the direction to the Sun, thereby causing elliptical orbits not far from circular. The analysis gave Kepler's first law of elliptical orbits for the planets, with the Sun at one focus. The law of planetary velocities that Newton derived was also in full agreement with Kepler's second law, the planet's line to the Sun tracing out equal areas in equal time. Kepler's first two laws had therefore been explained by simple algebraic equations, which allowed the calculation of the position and velocity of any planet at any time, past or future. Kepler's third law, that the square of the periods of the planets varied as the cube of their distance from the Sun, also came naturally out of the analysis.

Newton's theory went beyond the central problem of the motion of the planets, because it applied to the motion of all bodies within the Solar System. The orbits of comets were explained as bodies falling into the Solar System from outside. The orbit still has to be a conic section and, in this case, is a parabola, which is given by slicing a cone exactly parallel to one of its sides. But if a comet did not just fall into the Solar System but entered it with a significant inward velocity, then its orbit would be a hyperbola, which is the curve given by slicing a cone at an angle even more inclined than that parallel to one of its sides. Some comets, such as the famous Halley comet, have had their orbits perturbed by some planets, notably Jupiter, so as not to escape the Solar System; they become trapped with highly eccentric elliptical orbits. Comet Halley spends most of its life in the outer parts of the Solar System as defined by the orbit of the outermost planet, Pluto. Another great success of Newton's theory was the explanation of the tides as being attributable to the fact that the gravitational pull of the Moon and, to a lesser extent the Sun, was greater on the side of the Earth nearest to them, thereby causing the oceans to rise and fall in a periodic way, conditioned mainly by the Earth's rotation and the Moon's orbital period, but with variations in magnitude and timing, caused by the Sun.

In this work, accomplished mainly in his twenties, Newton had taken the findings of Galileo and extended them, he had developed a universal theory of gravitation, created a new mathematics and, from an entirely terrestrial basis, he had deduced the celestial laws of Kepler. Newton's cosmology not only described the Universe as it was known then but explained it as well. He had also united mathematics with nature, in the vision of Pythagoras more than two millennia previously.

Remarkably, Newton made no attempt to publish his immense achievement. He had worked alone and his insularity seemed to be such that he was quite happy to enjoy the fruits of his endeavours alone, seemingly content with the satisfaction of having successfully reached his goal of creating a fully consistent world model, and in this regard, he was totally different from most scientists. Fortunately, rumours of Newton's work spread and reached the ears of a young astronomer of good means, Edmond Halley, causing him to visit Newton in Cambridge. Halley quickly realized the epoch-making nature of Newton's work and persuaded him to publish it, offering to bear the full cost. Newton set about assembling his findings and preparing them for publication. This took him a few years because he had to make all of his analyses strictly rigorous, such as his instinctive assumption, which proved to be correct, that the gravitational effect of any body was the same as if its mass were concentrated in a point. He also had to clear up a few loose ends, but in 1687 his great work appeared under the title *Philosophiae Naturalis Principia Mathematica* (*Mathematical Principles of Natural Philosophy*). Like the great work of Copernicus, this was to become known by one word, the *Principia*. It was the foremost scientific treatise ever written and it became the foundation for a growth in science which has continued ever since. It brought Newton great fame and he was to be referred to freely as the greatest genius of all time, having many eulogies written about him by famous authors and poets. Gibbon was to write "The name of Newton raises the image of a profound genius, luminous and original"; a more perceptive Wordsworth was to refer to "Newton with his prism and silent face". There was also the following well known couplet of Alexander Pope:

Nature and Nature's Laws lay hid in Night;
God said "Let Newton be!" and all was Light.

Newton's interests extended well beyond his work on gravity and included alchemy, which he dabbled in without any kind of success for reasons we now understand. The transmutation of elements involve nuclear reactions vastly stronger than the chemical reactions that Newton was experimenting with, so much stronger that the atom was considered to be indivisible until it was finally split in the twentieth century. It is surprising that someone with the genius, insight and creativity to produce the *Principia* could have spent so much time in a completely blind alley. But this was the exception in Newton's work. He also conducted research in optics and made major seminal contributions to the study of light, which resulted in a second magnum opus simply called *Opticks*. Two aspects from this work are selected for discussion here because they represent the starting point for important developments in observational astronomy that are still continuing today.

First, Newton conceived, designed and built the first reflecting telescope, an instrument that still exists today and is held in the rooms of The Royal Society in London (see Colour Plate 4). All previous telescopes, including Galileo's, had been refracting telescopes, which used a lens that collected the light from the object being observed and then focused the light at a point where it could be viewed through another lens called the eyepiece. Newton realized that a concave mirror would also collect and focus light from an object, but since an observer's head would be in the way of direct viewing, he placed a second flat mirror inside the telescope tube to divert the beam sideways into the eyepiece. The development of astronomical telescopes has largely followed the reflecting route, which is more amenable to large sizes, but Newton's flat secondary mirror has been replaced by a curved convex mirror, which is not tilted and reflects the beam back down the telescope axis through a hole in the primary mirror where it can be viewed. This Cassegrain telescope (or its variations) is the basis for all of today's large astronomical telescopes, but, of course, the human eye has been replaced by a range of modern sophisticated detectors.

The second area of Newton's Opticks that had relevance for observational astronomy was his studies of the nature of light. He passed a beam of sunlight into a darkened room and through a glass prism, and the incident white light emerged as a coloured spectrum in which red light was refracted (deviated) least and violet light most; the order was red, orange, yellow, green, blue, indigo, violet (see Colour Plate 5). He eliminated the possibility that the colours were a property of the prism by passing the emerging spectrum into a second prism which was upside down compared to the first, and a single white beam emerged. Sunlight consisted of different colours which were revealed by the prism because it deflected them differently and thereby separated them. We now know that this is because light is a wave motion that can have different frequencies and wavelengths. The Sun emits over the whole range of wavelengths that the human eye can detect (and beyond) and the seven colours of the spectrum, seen in the rainbow, reflect the colour (or wavelength) resolution of the eye. After Newton's seminal research, instruments called spectrometers have been developed, which study the spectrum of the Sun and other celestial bodies such as the stars, with a spectral resolution orders of magnitude greater than the human eye alone. For reasons that I shall come to later, such spectroscopic observations have become one of the most powerful tools in astronomy, enabling the nature, composition and motions of the heavenly bodies to be determined. These major developments are of such importance that this branch of astronomy was given its own name: astrophysics.

One result of Newton's fame was to end the almost complete isolation with which he seemed content and had sought in Woolsthorpe and at Cambridge. He was brought into wider contact with the outside world, but unfortunately this led him into unseemly and quite unnecessary disputes with others as to the origins of his work. One argument was with Robert Hooke, a scientist of considerable distinction, who claimed that he had considered the inverse square law of gravity before Newton's *Principia*. Newton overreacted in responding at all because, as is clearly evident, no reaction was needed; his priority was secure, his position impregnable. A more serious and prolonged controversy was that with Leibnitz over the priority for creating the calculus. Again, this was unnecessary; it is clear that the two developments were quite independent and that Leibnitz created his calculus as a mathematical challenge, which he did superbly, and Newton invented his "fluxions" to apply his theory of gravitation to the Solar System, which he did superbly.

Newton's unitarian views disallowed him from being appointed as Master of a College, and this may have led to his leaving Cambridge in 1696 to become Master of the Royal Mint in London, a surprising post for someone of his particular talents, but one he was to hold for the rest of his life. More appropriately, he was elected President of the Royal Society in 1703 and knighted by Queen Anne in 1705. He died in 1727 and is buried in Westminster Abbey.

THE ERA OF
THE TELESCOPE

Part II deals with the remarkable extension of the human knowledge of the Universe that followed the exploitation of the telescope and the subsequent invention of the photographic plate and the introduction of astronomical spectroscopy. It covers the period from the end of the Renaissance to the middle of the twentieth century, a period that also saw the creation of a new natural philosophy with immense implications for astronomy.

CHAPTER SIX
The Classical
Post-Newtonian Period

Tales and golden histories
Of heaven and its mysteries
Keats

The period after the Renaissance did not suffer the dramatic decline in intellectual activities as that which followed the great age of the ancient Greeks. The Renaissance became the base and foundation from which all areas of science and scholarship grew and prospered. This was true of cultural activities in general, but this chapter will be confined to describing the progress in astronomy, together with those developments in natural philosophy that were relevant to astronomy. The period covered is from the seventeenth century to the very end of the nineteenth century, when a sea change was about to occur in both of these subjects.

Major advances were quickly made by the extensive exploitation of the telescope, which expanded the Universe accessible to human observation by a very great factor and whose power had been so dramatically demonstrated by Galileo. Observatories sprang up in the seventeenth, eighteenth and nineteenth centuries which were to make a major impact on astronomy. Notable among these were two of the earliest, the Observatoire de Paris, founded in 1671, and the Royal Observatory in Greenwich, founded in 1676. Each had different aims and functions, but both were to make great contributions to astronomy, as will be related shortly. In 1700 the Royal Observatory of Berlin was founded and, much later in that century, the private observatory of William Herschel, with its superb handmade telescope, was built in England and used by him to produce major astronomical returns. In 1825, another private observatory was built in Ireland by the third Earl of Rosse, with a 1.8 metres diameter, 18 metres long reflecting telescope, the largest in the world, but it did not produce the gains of Herschel's smaller telescope. In 1839, the Imperial Observatory of Russia was built in Pulkovo under the directorship of Friedrich von Struve, the first of a family line of distinguished astronomers. At the end of the nineteenth century, major observatories were founded for the first time in the USA; the Lick Observatory was built in 1888 on Mt Hamilton in California and the Yerkes Observatory at Williams Bay, Wisconsin in 1897. Both were based on refracting telescopes of apertures 91 centimetres and 102 centimetres, the latter still being the largest refractor in the world. They used multiple lenses of different shapes and different glasses so as to eliminate chromatic effects, and both were funded by donations from private donors whose names were given to each observatory. This funding of astronomy in America by private benefactors was to be an important factor in that country taking a leading international position in the twentieth century, a position that was heralded by the creation of the Lick and Yerkes observatories.

The first director of the Paris Observatory was Giovanni Demenico Cassini, an Italian who was Professor of Astronomy in the University of Bologna. He came to Paris with the intention of setting up the observatory and then returning to Bologna after a few years. In the event, he remained until his death in 1712 and was the forerunner of a remarkable astronomical dynasty, being succeeded as director by his son, grandson and great-grandson, all of whom continued the work that he started. His main interest lay with the planets; he determined the periods of rotation of Mars (24 hr 40 min) and of Jupiter (9 hr 56 min), and discovered four satellites of Saturn while also detecting the gap in that planet's rings which still bears his name. He correctly deduced that Saturn's rings were not continuous in nature but composed of small "moonlets". His other interest was in cartography, and he initiated a programme that was to be completed by his grandson about a century later with the final great map of France. These interests and experience suited him well for his most important contribution to astronomy – the first reasonably accurate determination of the size of the Solar System.

The first attempts to measure the distances to celestial bodies were made by the ancient Greek, Aristarchus, who obtained a reasonably accurate value for the distance to the Moon (about 80 per cent of its true value) but a hopelessly inaccurate value for the distance to the Sun (about 5 per cent of its true value). His method for determining the distance to the Sun was geometrically correct but depended on determining, with extremely high precision, the point when the Moon was exactly half full, a precision that was simply not possible then or now because of the irregularities on the Moon. It should be noted that these early estimates of distance were very much less than the actual values and, as will be seen in various parts of this book, such underestimates continued throughout the history of astronomy. At each stage in the development of new techniques to measure distance, the Sun, planets, stars and ultimately the external galaxies were revealed to be much farther away than previously estimated or thought, a reflection of the immense scale of the Universe compared with the Earth that we live on.

The next step in determining distance within the Solar System was taken by Kepler, who realized that an accurate measurement of the Earth–Sun distance was not possible at that time, although he did conclude that it must be at least three times greater than the estimate of Aristarchus. So, Kepler derived the distances of the planets from the Sun using the Earth–Sun distance as a baseline, and his measurements were therefore in these astronomical units. Consequently, his third law gave the relative distances of the planets from the Sun in terms of astronomical units and, hence, the measurement of the distance from the Earth to any other body in the Solar System would establish its scale and allow the magnitude of the astronomical unit, or Earth–Sun distance, to be determined. It was from this premise that Cassini developed his method – he would measure the distance to Mars.

As has been stressed earlier, any geometric distance measurement is more accurate when the baseline is largest and when the object under study is nearest. Cassini chose his object to be the planet Mars which was in opposition (closest approach) in 1673. For a baseline he chose the distance between Paris, from where he was to make the observations, and Cayenne in French Guiana, where the observations were to be made by his colleague, Jean Richer. The baseline was some several thousand kilometres long, and Cassini and Richer observed the exact position of Mars at the same time against the background of fixed stars. The different positions of Mars in angle allowed its distance to be determined from the

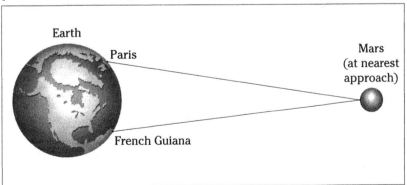

Cassini's measurement of the distance to Mars was based on simultaneous observations of the planet from two widely separated points on Earth: Paris and French Guiana.

simple method of triangulation used in surveying. The result, when translated into the Earth–Sun distance via Kepler's third law, was 140 million kilometres, only 7 per cent from the correct value of 150 million kilometres. The true scale of the Solar System had finally been established. It was staggeringly large, in terrestrial terms the distance to the Sun was about 12 000 times the diameter of the Earth, and the size of the Solar System as represented by the diameter of the orbit of Saturn, the most distant planet then known, was nearly a quarter of a million Earth diameters.

Another brilliant astronomer, who spent seven years at the Paris Observatory after its foundation, was a Dane, Olaus Romer, who was to return to Denmark as the Professor of Astronomy in Copenhagen where he remained for the rest of his life. But it was in Paris in 1675 that he made his great contribution, not only to astronomy, but to natural philosophy as well. In examining Cassini's extensive observations of Jupiter, he noticed that the duration of the eclipses of its moons (i.e. the time between the disappearance of the satellite behind Jupiter's disk and its re-emergence) varied in such a way that it was shortest when the Earth was approaching Jupiter at its highest velocity and longest when the Earth was receding at its highest velocity. The total change in eclipse period over a year, when the main change in relative velocity was attributable to the faster-moving Earth, came out to be about 22 minutes and Romer realized, rightly, that this was because the velocity of light was finite and not, as had previously been thought, infinite. When the Earth is approaching Jupiter, it is nearer to it when a satellite emerges than it was when it entered eclipse; the light therefore has a shorter distance to travel and, having a finite velocity, is seen earlier, causing a shorter eclipse duration. When the Earth is moving away from Jupiter, it is farther away when the satellite re-emerges and is therefore seen later, giving a longer duration. Hence, given that the distance from the Earth to Jupiter was known at the onset of eclipse and its emergence from eclipse, the velocity of light could be estimated from the time delays using simple geometry. The distances between the two planets could be calculated from Kepler's laws if the scale of the Solar System was known, and Romer was fortunate that this had just been established by Cassini in measuring the distance to Mars. He obtained a value of 214 000 kilometres per second compared to the true value of 300 000 kilometres per second. The errors involved were due to a combination of the error in Cassini's estimate of the astronomical unit, plus some errors because the motions of Jupiter's moons were not precisely regular. From then on, measurements of the velocity of light were to be based on terrestrial experiments, using methods optimized to match the kind of velocity established by Romer.

Whereas the work of the Paris Observatory seemed to be curiosity driven, the Royal Observatory in Greenwich, which was founded a few years later by Charles II and which was housed in buildings designed by Sir Christopher Wren, had a specific aim and purpose – to aid navigation so as to allow British sailing ships to travel the globe in aid of trade and, of course, war. This required the positions of stars to be measured accurately, together with those of the Sun, Moon and planets, so that their positions could be predicted accurately into the indefinite future. The programme was defined and initiated by the first Astronomer Royal, John Flamsteed, and resulted in a catalogue of about 3000 stars with positions accurate to about one second of arc. This was a major advance over the great catalogue of Tycho Brahe, who had recorded the positions of nearly 800 stars to an accuracy of about one minute of arc nearly 200 years before; the advance was made possible entirely by the application of the telescope. This work was extended to the Southern Hemisphere (British ships sailed everywhere) with the building of another Royal Observatory at the Cape of Good Hope in South Africa, founded in 1820.

The precise positions of stars allowed the latitude of ships to be determined accurately, but the determination of longitude was limited by the accuracy of the on-board clocks that were needed to tell the time in order to know the angular distance from the Greenwich meridian. The problem was one of such importance that the British government set up a Board of Longitude in 1714 and offered a prize of £20000 (an immense sum at that time) for a chronometer that would operate at sea and maintain a precision of a few seconds per day; this would enable longitude to be determined to better than 50 miles over the longest sea voyages, typically about six weeks in duration. A master clock-maker, John Harrison of London, met this requirement in 1764 (his second attempt) when his chronometer was successfully tested on a long sea voyage. However, he had to argue his case for many years and only received his full prize on the intervention of George III (presumably during a bout of sanity), to whom he appealed in 1773.

Telescopic studies of the planets, which were initiated by Galileo and greatly advanced by Cassini, received a large and unexpected boost when the five planets that had been known since ancient times were increased to six. This happened in 1781 when a German-born British musician and amateur astronomer, William Herschel, discovered a new planet with the aid of a superb reflecting telescope which he had built himself. He named the new planet Uranus after the mythological father of Saturn beyond whose orbit it revolved. Herschel was to join the professional world of astronomy that he had so surprised and impressed by his discovery, and he became one of the greatest observational astronomers since Tycho Brahe. His discovery caused intensive observations of Uranus to be made between 1790 and 1840 which revealed that its orbit was not perfect but subject to perturbations, which were proposed to be caused by gravitational interaction with another unknown and more distant planet. In 1845, John C. Adams in England and Urbain Leverrier in France independently calculated the orbit of the new planet and were able to predict its position. Although the observatories in their own countries were the first to receive these predictions, it was not until Leverrier passed the coordinates on to Berlin that Johann Galle discovered the object within one degree of the predicted position. All planets are named after Greek gods and this one was called Neptune.

To complete the story of the discovery of planets, the last and outermost planet was found in 1930 by Clyde Tombaugh at the Lovell Observatory in Arizona. Its discovery again followed predictions based on perturbations in the orbit of Uranus by using Newton's theory and it was named Pluto after the god of Hades. It has the most eccentric elliptical orbit of all the planets (at perihelium it falls within the orbit of Neptune) and has the greatest inclination ($17°$) to the ecliptic plane in which all other Solar System bodies revolve. Its discovery was a little fortunate because its large inclination was not predicted, but, at that time, it happened to lie close to the ecliptic plane where the search was concentrated.

The discovery of Neptune was one of the many great triumphs of Newton's theory of gravitation, but it is appropriate to mention here that there was another planet which displayed a very small deviation from a purely Newtonian orbit which could not be explained at that time in the nineteenth century. The planet was Mercury, the one nearest to the Sun, whose orbit was perfectly elliptical but whose perihelion, the point of its closest approach to the Sun, occurred not at the same identical position in each orbit but advanced very slowly in the direction of its motion. This advance was only 43 seconds of arc every century and its detection demonstrated the great precision that had become possible with telescopic astronomy. Like the non-Newtonian but different perturbations in the orbit of Uranus, the effect in Mercury was ascribed to another planet which would have to lie inside its orbit and closer to the Sun. Although calculations of the same nature as for Uranus were carried out, no inner planet could be found. The deviation was so small and Newton's theory so impregnable that no great attention was given to the effect, although the French astronomer, Urbain Leverrier, who, with John C. Adams, had calculated the position of the new planet Neptune, commented at the time that the effect must be attributable to "some as yet unknown action on which no light has been thrown". He was far more right in this statement than he could possibly have realized, as was to be revealed in the early twentieth century when Newton's theory of universal gravitation was to be superseded by another one of a totally different concept, developed by Albert Einstein and to be called "general relativity".

The great precision of astronomical observations that had become possible with the telescope, once again raised the hope in the eighteenth century that the parallax of stars, the slight shift in stellar position that should result from the motion of the Earth from one point in its orbit about the Sun to the diametrically opposite point that it would reach six months later, might finally be detected. This hope had been increased by the determination of the Earth–Sun distance from the observations of Mars by Cassini and Richer, because this was so much larger than had been thought hitherto; this afforded a much longer baseline and would therefore cause a larger angular shift for the same stellar distance. This renewed interest in detecting parallax was no longer cosmological, since the evidence in favour of the Copernican doctrine had become so overwhelming that it was no longer in scientific doubt. Hence, the detection of stellar parallax, which would previously have been undeniable proof that the Earth revolved around the Sun, was no longer necessary for that purpose, but it would allow the next major step in establishing the scale of the Universe by determining the distance to the nearest stars.

The first serious attempt was made by James Bradley, who was to become the third Astronomer Royal. Bradley had a telescope constructed by George Graham, a celebrated instrument-maker, and mounted precisely vertically in a fixed position so as to eliminate the bending effect of light caused by refraction when it passed through the Earth's atmosphere at an angle. He then selected the bright star gamma Draconis (the star labelled gamma in the constellation of Draconis) because this passed precisely overhead of London and any changes in its position relative to the vertical could be measured against a plumb line by micro-movements of an eyepiece. In 1728 he detected deviations that were greatest in one direction when the star was overhead just before sunrise and greatest in the opposite direction when, about six months later, it was overhead just after sunset. The magnitude of the shift was very appreciable, being about 20 seconds of arc in one direction and the same in the opposite. But the effect was later shown to be the same for any star in the same direction and therefore could not be attributable to stellar parallax, which is caused by the different *position* of the Earth on diametrically opposite parts of its orbit. The Bradley effect was due to the *velocity* of the Earth, which was in opposite directions on opposite parts of its orbit, and the displacement was always in the orthogonal direction of the Earth's motion relative to the star. Bradley had discovered a new effect to be called *stellar aberration*. It was caused by the fact that when the starlight entered the telescope, any transverse velocity resulted in the telescope moving slightly to one side during the brief period of time it took the light to pass along its axis; hence, to maintain the image on axis, the telescope had to be tilted slightly in the direction of the Earth's motion. The tilt was determined entirely by the transverse velocity of the Earth and the velocity of light and, unlike parallax, was totally independent of the distance of the star. For a star at the ecliptic pole, the direction of viewing is always orthogonal to the plane of the Earth's orbit, resulting in the Earth's velocity being always transverse to that direction and thereby causing the star to appear to move over one year in a circle of about 20 seconds of arc in radius. For a star in the ecliptic plane, the Earth's transverse motion varies from a maximum in one direction to a maximum in the opposite direction and the star will appear to move in a straight line from 20 seconds of arc in one direction and another 20 seconds of arc in the opposite direction, the whole motion again taking one year. For a star in between these two extremes, the apparent motion is an ellipse whose major axis is 40 seconds of arc, representing the maximum range of the Earth's velocity. The angle of aberration is given by the ratio of the Earth's transverse velocity to that of the velocity of light, and the maximum deviation of 20 seconds of arc as measured by Bradley was consistent both with the Earth's velocity (about 30 kilometres per second), as determined from Cassini's estimate of the Earth–Sun distance, and with the velocity of light derived by Romer.

Bradley's discovery and measurement of aberration was important because it afforded the first direct confirmation that the Earth was not stationary but actually was moving. This no longer had the cosmological and religious significance of more than a century earlier, but may have been one of the factors that led Pope Benedict XIV, in 1741, to remove the Catholic Church's ban on the writings of Galileo by having the Holy Office grant an imprimatur to the *Complete Works of Galileo*.

The astronomical importance of detecting stellar parallax still remained, but it had become very clear that its magnitude was very much smaller than that of stellar aberration or atmospheric refraction. The former was caused by the motion of the Earth and the latter by the bending of light by the Earth's atmosphere when starlight enters obliquely. But parallax has one property quite different from the other two, a property that presented the only hope of its detection; it depended only on the distance to the star observed, whereas the other two depended only on the direction. Hence, the detection of parallax depended on detecting the small change in the position of the nearer stars relative to the more distant background stars over a six-month period. It took about a century after Bradley before the precision was reached to detect this tiny effect, but in one year, 1838, it was achieved for three different stars by three independent observers. The first was Friedrich Bessel, Director of the Königsberg Observatory in Germany, who measured the parallax of the star 61 Cygni to be 0.3 seconds of arc. The second was Thomas Henderson, who measured the parallax of the bright southern star alpha Centauri from the British observatory at the Cape of Good Hope and obtained a value of nearly 0.8 seconds of arc.[1] The third person to detect parallax was the Russian astronomer Friedrich Struve, who used a superb refracting telescope built by the German instrument maker Joseph von Fraunhofer and installed in the observatory at Dorpat in Estonia. His star was alpha Lyrae (Vega) and the parallax was the smallest of the three at just over 0.1 seconds of arc, making it the most distant. Hence, in one year, the distances to three stars were determined from their measured parallaxes, using the astronomical unit as a baseline. But the distances were so vast by terrestrial standards that to express them in kilometres or miles would involve such huge numbers as to be incomprehensible. Even using the AU as a unit gave such large numbers that a new unit had to be found. The astronomers found one that came naturally from their method of measuring stellar distances – the *parsec*, which is a contraction of parallax second, the distance at which a star would have a parallax of one arc second. The parallaxes of all stars are less that one arc second, so that all stars are more distant than one parsec. For the three stars discussed above for which parallax was first measured, the distances in parsecs, starting with nearest, are 1.3 (alpha Centauri), 3.4 (61 Cygni) and 8.1 (alpha Lyrae). But the parsec is a somewhat technical unit, and a much better one, which has a more easily appreciated meaning and which will be used in this book, is the *light year*, the distance that light will travel in one year. This is a very large distance; in kilometres it is 10000000 million, in Earth diameters it is nearly 1000 million and in AU it is 60000. In these terms, alpha Centauri is about 4 light years away, 61 Cygni is 11 and alpha Lyrae is 26. Such immense distances to even the nearest stars were greeted by many with astonishment, but the measurement of parallax was to turn out to be the tiniest of steps on the road to establishing the scale of the Universe, as later chapters reveal.

The baseline for measuring stellar parallax is the astronomical unit, and the accuracy to which it was known therefore determined the accuracy to which the distance to stars could be determined. It had been determined by Cassini and Richer in 1673 by measuring the distance to Mars at its nearest approach to Earth, and in the eighteenth century considerable

---

1. It is of some human interest that Henderson made his observation a few years before Bessel, but illness made him delay the analysis and publication until after Bessel who, rightly, is accorded the priority of being the first to determine the distance to a star.

efforts were made to improve the accuracy. Since Kepler's laws established the relative motions of all planets in the Solar System, the distance to any one of them from Earth would establish the scale and hence the value of the AU. Cassini had selected Mars, but Edmund Halley, who had so encouraged Newton to publish the *Principia* and who became the second Astronomer Royal, proposed that Venus would be better because it was nearer to Earth and, despite the fact that at closest approach it would lie in between the Earth and Sun, it could be observed as a dark circle against the background Sun as it passed across the solar disk. Observations from different parts of the Earth, whose separations were known, of the precise times of appearance and disappearance of Venus would allow the AU to be measured. Venus is not precisely in the same plane as the Earth and does not always pass between it and the Sun, but it was known that transits would occur in 1761 and 1769, and many observatories in many parts of the world prepared for these events in an international effort to determine the size of the Solar System accurately. These included an expedition by Captain James Cook, the great navigator, to Tahiti to observe the Venus transit of 1769 in order to achieve wider global coverage of the event. But the accuracy of the method was constrained by effects caused by the outer atmospheres of the Sun and Venus, which limited the precision with which the planet could be located, and the attempts are now largely of historical interest.

The next advance in measuring the astronomical unit came a century later when, in 1877, Sir David Gill returned to the measurement of the distance of Mars which, because of the eccentricity of its orbit, was at its nearest possible approach to Earth. But he improved on earlier measurements by using the same telescope and observing the location of Mars against the background stars just after sunset and just before sunrise. This gave a baseline that was a significant fraction of the Earth's diameter and, by using the same telescope and the newly introduced photographic plate, other errors were minimized. Of course the Earth and Mars had moved in their orbits between observations, but this could be allowed for because such movements were known precisely in terms of the quantity being measured, the astronomical unit. Since Mars is a disk, the accuracy to which it can be located is a little limited and this led Gill to propose that the same method should be applied to the minor planets, which had then been recently discovered between the Earth and Mars since these would appear point-like and were closer than Mars. He conducted these measurements from the Cape Observatory in South Africa during 1888 and 1889 when the three minor planets, Victoria, Isis and Sappho, were at their nearest approach; he obtained a very accurate estimate of the astronomical unit which lies within 0.1 per cent of today's precise value of 150 million kilometres, achieved by radar reflections from the Sun and planets. In light travel terms, the astronomical unit is 500 light seconds and it is the baseline for determining the distance to the stars and ultimately the size of the Universe.

The astronomical studies of the eighteenth and nineteenth centuries were mainly, but not entirely, devoted to the stars and planets. A French astronomer, Charles Messier, had a great interest in comets and scoured the sky in search of new ones. But there were other "nebulous" objects in the sky which he knew were not comets because they did not move relative to the background of fixed stars and these diverted him from his task because it took him some time to establish their non-movement. Accordingly, he decided to record and catalogue these so that he could recognize them and quickly discard them as cometary candidates. In 1784 he published a catalogue containing 109 of the brightest "nebulous" objects, which are still called nebulae today, and each one is still referred to by the number it is given in his catalogue; for example, the first object, which is the Crab Nebula (already mentioned above) is designated as M1. At that same time, the great British astronomer, William Herschel (discoverer of Uranus) embarked on a far more exhaustive survey of nebulae which was carried on by his son John and extended to cover the Southern Hemisphere as well as the northern by using the observatory at the Cape of Good Hope. The Herschels recorded 7840 objects and these were later published and extended by J. L. E. Dreyer at Armargh Observatory in the *New General Catalogue of Nebulae* to which he later added two supplements. Like the Messier catalogue, objects in the *New General Catalogue* are also assigned numbers, still used today, so those 109 objects that are common to each catalogue have two numbers; for example M1, the Crab Nebula, is also NGC 1952. The catalogues of nebulae were of great astronomical importance and the study of the objects listed therein was to continue into the twentieth century. Their nature varied enormously, even though most are still called nebulae, and they were given descriptive names according to their appearance (e.g. diffuse nebulae, planetary nebulae and spiral nebulae). Some were the sites of recent star formation where the new young hot stars caused the medium they condensed from to fluoresce in their ultraviolet radiation fields; others were older stars which had shed their atmospheres, sometimes cataclysmically, in the late stages of their evolution; some were clusters of stars that had appeared nebulous to Messier and the Herschels; and some were a variety of external galaxies. None of this was known at the time and the identification of the nature and astronomical significance of the nebulae will be explained later at the appropriate point in the telling of this story.

The very extensive telescopic observations of the sky caused a totally different view of cosmology to that of Aristotle, who had pictured the outer Universe as a thin spherical shell of stars surrounding the Earth, Sun and planets. This view had already been severely weakened by Galileo, whose telescope revealed large numbers of stars too faint to be visible with the naked eye, implying that they lay at greater distances. As the quality and size of telescopes increased, more and more stars were always revealed, leading to the view that the Universe was infinite and composed mainly of stars because these were far more numerous than the nebulae. Although this was very different from Aristotle's spherical shell, it was hardly more sophisticated, but it did stimulate a German astronomer, Heinrich Wilhelm Olbers, to place, for the very first time, constraints on the nature of the Universe at large. He did this by considering the implication of an astronomical observation that every reader has made and knows about; indeed it is a major factor in everyday life – it gets dark at night! This led him to ask a question which has become known as Olbers' paradox: why is the sky dark at night? In 1826, Olbers pointed out that, if the Universe was infinite and composed of stars, then every line of sight from the Earth would eventually coincide with a star and the sky would be brilliantly bright. If all stars were like the Sun, the sky would be everywhere as bright as the surface of the Sun. (Of course if it were that bright there would be no life on Earth, and probably no Earth, and the question could not have been asked.) Olbers concluded that we must be living in a Universe that was finite in either space or time. The latter was a possible explanation because of the finite velocity of light; hence, a spatially infinite Universe, which had been formed a finite time ago, could only be seen as far as light could travel in that time and would therefore appear finite. From now on, all theories of the Universe, all cosmologies, had to be able to answer Olbers' paradox. But although both the argument and conclusion of Olbers were strictly correct, they were based on the assumption that the Universe is static, an assumption that was not regarded as such because it had been accepted with conviction since ancient times. Indeed, the conviction of a static Universe continued in astronomy for a century after Olbers published his paradox.

In the middle of the nineteenth century, the telescope, which was being continually increased in size and quality, was joined by two other technical developments, also of epoch making importance. The first was the introduction of the photographic plate, the first man-made detector which, with its ability to integrate over time, was able to outperform the human eye by far and, as a major bonus, left a permanent record of the observation. Much fainter objects than had been seen hitherto were observed, with a great increase in the number of stars and nebulae recorded. The great density of stars in the Milky Way became even more apparent and the first crude attempts to measure the distance to the most distant stars were embarked upon by measuring their brightness and assuming that their faintness compared to the Sun was attributable to their distance. These first attempts were highly inaccurate and, like all early attempts to measure distance in astronomy, were gross underestimates. Nevertheless, they gave a picture of the Universe which was a definite step forwards – a flattened Universe of stars lying in the plane of the Milky Way, which was a few thousand light years across. How inaccurate this was will be revealed later, but for the moment, towards the end of the nineteenth century, it answered Olbers' paradox: the Universe was finite (although staggeringly large by terrestrial standards) and had the shape of a flattened disk with the Sun near its centre. Outside of the Universe there was nothing, a complete void.

The second major development in observational astronomy was the introduction of the spectrometer which, attached to the end of a telescope, would spread the light from a celestial body into its different colours (wavelengths) usually, following Newton, by means of a prism. The spread of light diminished its intensity at each point below the detection limit of the human eye, and the full application of the spectrometer was dependent on the use of a photographic plate; the spectrometer was therefore usually referred to as a spectrograph. The first spectroscopic observations of stars were made by the English astronomer, William Huggins, in 1864 and these were followed by many others, notably the great lady astronomer Annie Cannon, who carried out a major programme at Harvard College Observatory in New England, which resulted in more than half a million spectra of stars being recorded. She noted the great variation in the nature of the spectra of stars and classified these into seven different categories, into one of which each star was assigned. The resulting *Henry Draper Catalogue* (named after one of those private benefactors who generously funded American astronomy in the late nineteenth century and who were to continue into the twentieth century) formed the starting point for the study of the nature and evolution of stars, a story to be told later.

The first astronomical spectrum to be observed was that of the Sun which was bright enough to be seen easily with the naked eye. This was studied by the German scientist, Joseph von Fraunhofer, in the early nineteenth century and he noted and recorded the presence of many dark lines in the otherwise continuous emission spectrum of the Sun. Parallel studies were being made in the laboratory of the spectra of different elements by isolating each element in a gaseous form in a discharge tube and causing it to become luminous by passing an electric current through it. The remarkable property was discovered that all elements emit only in discrete, fixed wavelengths (or colours) and that these are different for every element. That means that the spectrum of each atom is unique to that atom, like a human fingerprint, and its presence in any source can be unambiguously determined from its spectrum. The spectral lines emitted by the discharge tube method appear bright, but if the same element sample is placed in front of a very bright incandescent source (such as a white-hot filament), the same spectral lines are observed as dark lines in the continuous spectrum of the background source. Hence, each atom can emit or absorb the same unique spectrum. In the case of the Sun, it was the latter effect that was being observed and the dark lines are now called Fraunhofer lines after their discoverer. A major astronomical step could now be taken. Many of the Fraunhofer lines corresponded to spectral lines of elements observed in the laboratory, thereby establishing the existence in the Sun of elements present on Earth; these included hydrogen, calcium, sodium and magnesium, a list greatly extended since then.

The great success in identifying the dark absorption lines in the solar spectrum, with the spectra of elements seen in the laboratory, was not repeated when spectral studies were made during total solar eclipses. These rare events are caused by the astronomically fortunate chance that the angular diameters of the Moon and Sun, as seen from the Earth, are almost identical, so that, on those occasions when the Moon passes exactly between the Sun and Earth, it blots out the whole of the bright solar surface, which is called the photosphere, because it is the source of most of the light. These solar eclipses revealed the presence of a very extended medium, stretching out to a distance greater than a solar radius, which became a matter for intensive study in the second part of the nineteenth century. Observations revealed a bright inner portion, called the chromosphere because it is coloured, and a very extensive fainter halo, called the corona because it completely surrounds the Sun. In contrast to the dark absorption lines in the photosphere, spectroscopic studies of the chromosphere and corona revealed lines that were bright and in emission; the reason for this was not understood at the time but we now know it is because the chromosphere and corona, although much more tenuous than the photosphere, are very much hotter because of heating processes not yet fully understood. But the more important point was that the wavelengths of the strongest emission lines in the chromosphere and in the corona could not be matched to the lines of any element that had been observed in the laboratory at that time. The proposed solution to this problem, made by the British astronomer Norman Lockyer, was that the eclipse spectra were explained by elements as yet undiscovered on Earth and that lines in the chromosphere were emitted by the unknown element helium (named after the Sun) and that lines in the corona were emitted by another unknown element called coronium. The proposal was correct in the first instance, although helium was not discovered on Earth until nearly 30 years later in 1895, when it was chemically isolated by Sir William Ramsay at University College London. In the second case, the proposal was completely wrong, but a further 50 years had to elapse before the spectral lines emitted by the solar corona could be identified as being attributable to well known elements but in a very unexpected state, as will be related later.

With the advent of stellar spectroscopy, the search for terrestrial elements was extended from the Sun to the stars. Again it was found that many of the prominent lines present in the spectra of stars could be identified as being caused by elements present on Earth. The stars, as well as the Sun, were made of familiar materials. A common chemistry had been established in the Universe and this momentous result heralded the great importance of spectroscopy, which introduced a new technique into astronomy, to be given its own name of astrophysics, and which has developed into the most powerful tool for investigating the detailed nature of all celestial bodies and of the Universe itself.

Developments in natural philosophy (or physics, as it became to be called) have always been pertinent to the study of astronomy. That was true in Newton's time and it is still true today. Major progress in physics was made in the eighteenth and nineteenth centuries, much of it relevant to astronomy, such as the discovery, outlined above, that each atom has its own unique spectrum or fingerprint. Other relevant developments will now be discussed.

Physics is the study of energy and its interaction with matter. Energy exists in a wide variety of forms, and any changes from one energy form to another always involve matter and are conditioned by the *laws of thermodynamics*, which are among the most fundamental laws of nature. They are therefore of great importance to astronomy and we shall return to them later in a cosmological setting. The laws were compiled in the first half of the nineteenth century and have since stood up to every experimental test to which they have been subjected. The first law of thermodynamics states, very simply, that energy is conserved; that is, that energy can be transformed from one form to another, but the total energy in any system and, indeed, the Universe, cannot be increased or decreased; it can be exchanged but it cannot be created or destroyed. It has its counterpart and close parallel in the *law of conservation of matter*, which says the same thing about matter as the first law of thermodynamics says about energy. Matter can be changed from one form to another, but the total amount of matter is fixed and is based on the indestructible atom, which can be formed into a wide range of diverse molecules and materials but which itself cannot be created or destroyed. In the early part of the twentieth century, these two laws were to become inherently linked and placed in a wider, more generalized setting.

The second law of thermodynamics can be stated as simply as the first, but it requires greater explanation. It states that the entropy of any system (including the Universe) will always increase with time until it reaches a maximum (or a stable plateau). Entropy can be regarded as a measure of the amount of energy in an isolated system which is no longer capable of change; when it reaches its maximum, the system is in its final stage of equilibrium. The other energy in the system is still capable of change and is still "useful", but when entropy reaches its maximum, there is no useful energy left. As an example, let us consider the simple but familiar example of a room that needs to be heated and contains a fireplace in which there is a grate filled with coal. The entropy of the room, its furniture, air and walls, has maximized, but that of the coal is at a stable plateau and it still has useful energy to be released by raising its temperature enough for it to combine with oxygen, in other words to burn. The coal is lit and the energy released heats the room, but when the coal is exhausted, it cools, and its ash and the room all reach the same temperature when the entropy has reached a new maximum. By the first law, the energy released by the burning coal still exists and is in the form of heat, but it cannot be used again, for example to heat a kettle. However, it can be transformed into some other form, for example by extracting from the room and depositing in some other place, say outside, in a refrigerating mode, but this requires the availability of some extra usable energy, and the entropy of the total system, room plus refrigerator, still increases.

The rate at which energy changes occur depends on each individual situation, such as the rate by which a hot body cools by transmitting its heat to the surroundings; the second law of thermodynamics simply indicates the direction in which such energy transformations flow. As such, it is a law that defines the direction of time. The physical meaning and understanding of the second law of thermodynamics was greatly advanced in the second half of the nineteenth century by the Austrian physicist, Ludwig Boltzmann, who, with great insight, recognized that entropy could be identified with disorder and that the ultimate state of any system would be its most random and therefore its most probable. He was then able to express the entropy of any system in terms of the probability that any individual state would occur, and this involved a universal constant which now bears his name. He applied this to the kinetic theory of gases, which were then considered to be composed of atoms or molecules that were free of each other but were in constant elastic collision, their velocities representing the temperature of the gas, which increased as the velocities increased, and whose energy represented heat. The velocity of any one molecule

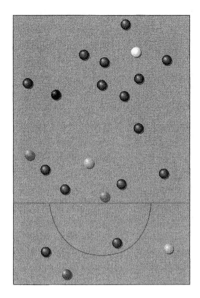

One of the fundamental laws of nature is the second law of thermodynamics, which states that as energy transformations occur, less and less energy will be available to do useful work. It also drives systems from orderly states to random states and, as such, indicates the direction of the arrow of time. The above two representations are of a snooker game and the reader will have no problem in identifying which was earlier and which was later.

changed continually at every collision but, overall, the average distribution of velocities was very constant, because of the very large number of molecules, and was determined by the entropy of the whole system reaching a maximum. By calculating the probability that any molecule had a certain velocity, it was possible to determine the distribution of velocities of molecules in a gas at any given temperature. Heat was now understood. The distribution showed a peak at the velocity that most molecules had, and the number of molecules with lower velocities steadily decreases, as do the number that have higher velocities. To give a feel for the quantities involved, let us take air at room temperature. The mean velocity of the molecules is such that they would travel a third of a metre in a millisecond; this velocity is also the speed at which the air can transmit noise, giving the velocity of sound in the more familiar units of 1200 kilometres per hour. However, the air molecules do not move in uninterrupted straight lines, but suffer, on average, several million collisions every second in which their velocities, as well as directions, change abruptly. But the overall distribution, averaged over all molecules, remains the same.

The second law was to be used in the mid-twentieth century by the scientist and successful novelist, C. P. Snow, as one example of his concern at the intellectual separation that had developed between science and the humanities, a separation that he had identified in his controversial book *Two Cultures*. He said in a speech to the House of Lords that anyone who did not know the second law of thermodynamics could not consider him or herself to be educated. This somewhat provocative remark was probably made with a tongue in the cheek, but since this book is largely aimed at the community beyond the physical sciences to whom Snow's comment was directed, I would hope that, having read this far, you can now meet the Snow criterion! But although most readers will not have known the second law of thermodynamics in the formal scientific terms in which it is expressed in, many, if not all, will have recognized it from their everyday experience; the coal in the grate, the oil in the tank, the petrol in the car has to be replaced when used, and a switch has to be thrown in order to draw on more electricity, whose use is then carefully monitored for payment. Further, since human beings (like all other forms of life) are also subjected to the second law of thermodynamics, they have to replace their energy as it is used and do so by the consumption of food which is needed every day. The cosmological implications of the laws of thermodynamics are as profound as those which arose from the fact that the sky is dark at night. Since the total energy of the Universe is conserved by the first law, and since the total entropy of the Universe is increasing by the second law, the amount of energy that still has the ability to change to some other form is diminishing inexorably and will inevitably become zero. The Universe is running down and, ultimately, the stars and galaxies will no longer shine, and it will effectively be a non-changing dead world. But there is a long, a very long, way to go before then.

Of great relevance to astronomy were the laboratory studies of the nature of light, and a major step forwards was made in the early nineteenth century when Thomas Young showed that light was some form of wave motion by demonstrating that it had the properties of diffraction and interference. He did this with a very simple experiment, in which he placed a single light source behind a screen which had two closely separated parallel slits cut into it. Behind the screen he observed not two bright lines produced by the slits but a series of fringes – bright regions interspersed with dark regions. The light beams emerging from the two slits had spread out (diffracted) because of their wave nature, thus allowing them to overlap and, in overlapping, their wave nature caused them to interfere, because the path lengths of the two beams were slightly different and their electromagnetic oscillations were no longer in phase. For a path difference of zero, the oscillations were still in phase and the two beams enhanced each other to give a bright fringe, but when the path difference became half of the wavelength of the light they were exactly out of phase and destroyed each other to give a dark fringe. When the path difference became one whole wavelength, they were once again in unison to give a second bright fringe, and so on. Different colours produced differently spaced fringes, demonstrating that colour was caused by variations in the wavelength of light. Bright fringes occur when the path difference is an integral number (0, 1, 2, 3, . . .) of wavelengths and, since the path difference could be measured from the geometry of the system, the wavelength of light could be measured and hence its frequency could be derived from the velocity of light, since this is equal to the wavelength multiplied by the frequency. To give a feel for the quantities involved, let us take light that is seen as green by the human eye: its wavelength is very small at half a micron (one micron equals a millionth of a metre); its frequency is extremely high, since it oscillates 600 million million times a second.

Since light is the result of a wave motion, it was realized that its wavelength, and hence its frequency, would appear modified to an observer for whom the source was in relative motion. If the source were moving towards the observer, the waves would squash up and the wavelengths would appear shorter; if the source were moving away from the observer, the wavelengths would be stretched out and would appear longer. In the former case, visible light would shift in the direction of the blue end of the spectrum and, in the latter, it would shift towards the red; these are therefore known as a blueshift or a redshift. These findings were made by the Austrian physicist, C. J. Doppler, in 1842 and the effect is known as a Doppler effect, and the shift as a Doppler shift. The magnitude of the shift in wavelength as a percentage of its wavelength is equal to its velocity as a percentage of the velocity of light. This presented astronomy with a technique of great importance; by observing the spectra of moving celestial objects and comparing the wavelengths of spectral lines due to different elements with the wavelengths of the same elements produced in a stationary discharge tube, the velocity of the object relative to the Earth can be determined. A redshift means that the celestial body is receding, and a blueshift means it is approaching; since the speed deduced is that which is in the line of sight to the Earth, the velocity is always called the *radial velocity*. Its application to astronomy has allowed the dynamics of stars, galaxies and even the Universe to be determined.

Until the early seventeenth century, the velocity of light was thought to be infinite and therefore any observed phenomenon would be seen instantaneously, but this was shown to be incorrect (as related earlier) by Olaus Romer when he noticed that the time intervals between the disappearance of any of Jupiter's moons behind its disk, and their reappearance,

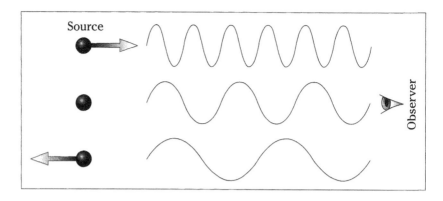

Source

Observer

The Doppler effect is a very important phenomenon in astronomy and is represented here. A movement of a source compared to an observer either squeezes up (blueshift) or stretches out (redshift) the light waves and allows the velocity and direction of the motion to be determined.

varied according to whether the Earth was moving towards or away from Jupiter. He correctly concluded that this was because the Earth was closer on re-emergence than it was on disappearance in the first case and farther away in the second case. This enabled him to make the first estimate of the velocity of light but, since the method suffered from some inherent inaccuracies, attempts to measure it moved into the laboratory. The very high velocity involved (300000 kilometres per second) posed severe experimental problems and it was not until 1849 that the first successful experiment was conducted by Armand Fizeau. He developed a rapidly rotating cog-wheel and placed this on a mountain top in front of a very bright light source, thereby causing it to flash rapidly on and off. On another mountain some 8.6 kilometres away he placed a large plane mirror which reflected the light back to the same cog-wheel. The returning light signal had a time delay caused by the time it took to reach the reflecting mirror and return. When the speed of rotation of the wheel reached the point where the time delay was the same as the time to move by half the separation of each cog, the return beam was blocked, and when the speed reached the point that the wheel had moved a full cog separation during the time delay, the return beam could be viewed again. This enabled the time of flight of the light beam to be measured, and from the distance, the velocity of light could be determined.

This method was quickly improved by Jean Foucault, who replaced the cog-wheel by a rapidly rotating mirror which reflected a source towards a distant plane mirror. On reflection from the plane mirror the rotating mirror will have turned to a slightly different angle, depending on its speed of rotation and the time of flight, and its image will therefore be displaced compared to a non-rotating mirror. Towards the end of the nineteenth century, this method was further refined by one of the first great American scientists, Albert Michelson, who obtained a value within one-tenth of a per cent of today's value, which has been obtained by the use of precise electronic timing units. Michelson was to extend his studies of light by developing the interferometer named after him, in which he removed half of the light from a source with a partially silvered mirror, took it on a different path and recombined it with the original beam, thereby creating a huge path difference in terms of the wavelength of light. This resulted in an instrument of immense power for measuring the properties of light. With his colleague, E. W. Morley, he used this instrument to conduct an experiment which, although it had a null result, had immense implications and was to be regarded as one of the most important experiments of the nineteenth century. The Michelson–Morley experiment was an attempt to measure the change in the velocity of light caused by the Earth's motion. Since light was a wave motion of some kind, it had to be transmitted, like sound, through a medium which was called the ether. Its velocity therefore depended on the relative motion of the ether and the observer; in the analogy with sound, the velocity depends on whether there is a wind or whether the observer is moving with respect to the air. The velocity of the Earth, at 30 kilometres a second, is 10 000 times less than that of light, but the Michelson interferometer had the power to measure this. The null result meant that the velocity of light was constant and did not change whatever the direction or magnitude of the velocity of the Earth! This seemingly impossible result was regarded as incredulous in many scientific circles and it tended to be disregarded as erroneous. Early in the next century, its significance was to reach mammoth proportions and it would become a part of the greatest revolution natural philosophy had experienced since Isaac Newton.

At the time of Newton one fundamental force of nature that was understood and quantified was gravitation, but it was realized that at least one other force of nature must exist – that which held matter together. This was best expressed by Newton himself who said, with a quotation that adorned the entrance to the British Rutherford Laboratory when it was set up to investigate the fundamental forces of matter, that:

There are therefore agents in nature able to make the particles of bodies stick together by very strong attractions, and it is the business of experimental philosophy to find them out.

In the post-Newton era, two other forms of energy of a more mysterious kind were being studied, electricity and magnetism, but these studies were hampered by the fact that the former could not be produced in any controlled fashion, being the result of processes such as the frictional interaction between two materials, and the latter was available only in the natural properties of magnetic material. This situation was changed dramatically by the achievements of the Italian scientist, Count Alessandro Volta, who made the first electrical battery in 1800, using zinc and copper plates. This gave a controllable source of electricity which stimulated, as new technologies always do, scientific studies that led to major advances. Of these, the most outstanding were those made by Michael Faraday at the Royal Institution in London. Faraday was born in 1791 into a family of very modest means and had a very limited education. Nevertheless, he had an innate genius which gave him deep insight into natural phenomena and which was to lead him to become one of the greatest experimental philosophers of all time. Of the great scientists referred to in any part of this book, Faraday is probably the nicest and certainly the most noble. He avoided social advancement and the pursuit of personal wealth, even though many of his discoveries, if patented, had immense commercial value. His refusal of high honours, including the presidency of the Royal Society (the greatest accolade for any British scientist), was caused by his considerable reserve and natural modesty. In a letter explaining this, he said "I must remain plain Michael Faraday to the end".

At the age of 14, Faraday was apprenticed to a bookseller and binder and he was to continue his training for eight years, during which his innate interest in science caused him to read some of the relevant books he dealt with, particularly those dealing with the new and mysterious phenomenon of electricity. As can so often happen in life, a chance event led to a major change in the nature and direction of his career. An appreciative customer gave him free tickets for a series of lectures on chemistry to be given at the Royal Institution by its then director Sir Humphrey Davy (always now referred to as the inventor of the miner's safety lamp). Faraday was entranced by the lectures, wrote them up in very neat writing, had them bound and sent them to Davy with a request for employment at the Royal Institution. Davy was very impressed and appointed Faraday as his laboratory assistant in 1813, when Europe was in the closing stages of the Napoleonic Wars, and when Faraday was 22. Even though he did not have the academic qualifications required to enter scientific research in his own right, he was ultimately to succeed Davy and become the most illustrious director that the Royal Institution has ever had. He made many contributions to physics and chemistry, but only those relevant to this book will be described here; they happen to represent his greatest achievements, namely his investigations into the nature of electricity and magnetism. He started these in 1820, when he was 29, and, after studies in many other areas, he returned to them in 1830. The whole programme represents a series of some of the most brilliant experiments ever conducted, in which he demonstrated that magnetism was the result of a moving electric charge, in other words an electric current. Hence, the two forces were not distinct; Faraday had unified them into one force, electromagnetism, the force that, as Newton had said, makes "bodies stick together". There were now two known forces of nature – gravity and electromagnetism – but the latter had a long, long way to go before it could be explained and quantified like the former.

The background to these experiments was interesting and possibly informative to the policy being applied today for the funding of science in many countries. Faraday had been working for some time on a task aimed at improving the quality of glass, which was funded by an external sponsor for strictly commercial reasons. Faraday made considerable progress, but after a while his thoughts would stray to electricity and magnetism and he began to feel slightly jaded; accordingly, he wrote to his sponsor, asking if he could be allowed a break in order to pursue his other interests and thereby to rejuvenate himself. During that break he invented the electric motor, and later added the electric dynamo and the electric transformer, thereby establishing the basis of the modern electricity industry with such immense practical and economic value that it has transformed human life in modern society.

Faraday showed that a magnetic field was the result of an electric current, and had a strength that increased with the magnitude of the current and a direction orthogonal to it; hence a straight wire carrying an electric current produced a circular magnetic around it, and a current through a coil produced a straight magnetic field along the axis of the coil. Since some materials were naturally magnetic (such as magnetite, a form of iron oxide) this implied that they contained microscopic circular currents which happened to be lined up with each other; this was not to be understood until the following century when the electrical and magnetic properties of the atom were revealed. Faraday then proceeded to investigate the interaction of electric currents, with their self-induced magnetic fields, with other electrical and magnetic configurations. One of his first experiments (conducted during the break he had sought to seek rejuvenation) was very simple but also very elegant. He suspended a needle from a point about which it could freely swivel, and the end of the needle was dipped into a bowl of mercury (an electrically conducting liquid). He placed this system in a transverse permanent magnetic field and then passed an electric current through the needle from a battery. The needle tilted and then rotated around its pivot, only stopping when the current was switched off; if he reversed the current, the needle again rotated through the mercury bath, but in the opposite direction. Faraday had constructed the first electric motor. He was later to demonstrate the reverse process by rapidly inserting, by hand, a bar magnet into a wire coil, in which he induced an electric current: the first dynamo. He refined this by making a copper disk which could be rotated between the poles of a permanent magnet, thereby generating a constant current. Subsequent developments of electrical generators were to use the more efficient method of rotating coils rather than disks, but since the direction of the plane of the coil relative to the magnetic field reversed itself in every cycle, the output current also reversed; this explains why our modern-day electricity supply is in the form of an alternating current.

Finally, Faraday discovered the phenomenon of magnetic induction, by which energy can be transformed from one electrical circuit to another without any physical contact whatsoever. He passed a current through a metal coil, which was placed close to another separate coil in which a current was induced, even though there was no physical contact between the two; he further found that the voltage induced in the secondary coil could be increased or decreased according to the relative number of turns in each coil: he had assembled the first transformer. Of course, any increase in voltage is accompanied by a decrease in current (and vice versa) because the total energy is given by the product of these two, and this is conserved by the first law of thermodynamics. Inevitably, as demanded by the second law of thermodynamics, some useful energy will be lost and this appears in the form of heat in the two coils.

The scientific and commercial consequences of Faraday's researches into electromagnetism were quite immense, but his interest lay only in the former. In order to illustrate this aspect of his nature and to demonstrate his own view of the great advances he made in natural philosophy, it is appropriate to defer to his own statement: "The great beauty of our science [is] that advancement in it . . . opens up the door to further, abundant knowledge overflowing with beauty and utility".

Faraday's great discoveries inspired James Clerk Maxwell (1831–79) to his equally great achievement of developing his electromagnetic theory of light as the propagation of an oscillating electric field, with its attendant magnetic field oscillating at right angles. In this brilliant theory, Maxwell derived an expression for the velocity of the electromagnetic waves which was independent of their frequency or wavelength, and given in terms of electrostatic and magnetic quantities that had been measured in the laboratory. These were quite unrelated to the phenomenon of light itself, but gave a value for its velocity that agreed with the velocity that had been measured directly at that time, providing overwhelming proof of the theory itself. Maxwell's electromagnetic theory of light was completed in 1871, a year that saw the proclamation of a united Germany and the publication of Darwin's second great work on evolution, the *Descent of Man*.

Until that time, studies of light had meant studies of visible light, and these were carried out using incandescent sources in which the basic emission was caused by atoms excited by the particular heating or discharge processes. Maxwell's theory clearly implied that the ultra-high frequencies involved in visible light indicated that atoms had to have an electrical property that could oscillate at such frequencies; the understanding of this had to wait until the twentieth century. But Maxwell's theory also showed that an electrical oscillation of any frequency would produce an electromagnetic wave of the same frequency which would propagate at the same velocity, namely the velocity of light. This was a specific prediction which was put to the test by a German scientist, Heinrich Hertz, in 1888. Hertz built two large conducting plates, each with a protruding wire, which were adjusted so that they were separated by a small gap. He then charged one plate positively and the other negatively by applying a high voltage between them. When the voltage became high enough, a spark occurred by breakdown in the small gap, electricity flowed from one plate to the other until the charge difference was reversed, when the spark recurred because of the electric flow in the opposite direction. He had produced the first man-made electrical oscillator whose frequency was almost 100 million per second. A simple detector, consisting of a circular loop with a small gap in it, was placed at the other end of his laboratory, and a current was induced in this, causing it to spark when the oscillator was fired. But this was not enough to prove the existence of an electromagnetic wave emitted by the oscillator, since it could be some form of Faraday-type induction, so he placed a radio reflector behind the detector which would reflect back any electromagnetic wave. The detector was therefore subjected to the direct wave from the oscillator and the reflected one from the mirror, and, since that had travelled an extra distance, there was a path difference between them and interference would occur, as had been demonstrated with visible light. By moving his mirror in the direction of the oscillator, Hertz detected the interference by seeing a high signal when the waves were in phase and a zero signal when they were in anti-phase; on continuing the movement of the mirror, the intense signal returned and the waves were once more in phase. The total change in path length from one maximum to the next was equal to a wavelength that Hertz was then able to measure easily; he obtained 5 metres, which when multiplied by the frequency of his oscillator, gave a value for the propagation velocity that was in agreement with the measured velocity of visible light. Hertz had fully confirmed Maxwell's theory and in so doing had discovered radio waves.

Only a few years after the experiments of Hertz, the German physicist

Wilhelm Konrad von Röntgen, discovered X-rays by driving very high voltage discharges into solid targets. He quickly realized that the radiation he was generating had a great power of penetration and he demonstrated this with the very first X-ray photograph, of his wife's hand, and clearly showed her bones and their structure. X-rays are also electromagnetic waves but of a very much higher frequency and therefore much shorter wavelength than visible light.

At about the same time, in the closing years of the nineteenth century, the French physicist, A. H. Becquerel, detected a new form of radiation which emanated from compounds of uranium and was sufficiently penetrating to expose photographic plates held in completely light-tight cartons. Von Röntgen had just discovered X-rays in a similar fashion, but these were present only when his high voltage discharge tubes were switched on. In the case of a uranium compound, the rays were present all the time and were being emitted passively and spontaneously of its own accord without any external stimulation: Becquerel had discovered radioactivity. Many scientists embarked on studies of the properties of this new and mysterious phenomenon, prominent among whom were Pierre and Marie Curie, working at the Sorbonne in Paris. Pierre was the professor of physics and his wife, the Polish-born Marie, worked with him and was one of the first truly great women scientists whose entry into a male-dominated field was achieved by marriage. She was to prove to be more brilliant even than her husband and she received a Nobel prize in her own right in 1911. The Curies established that the intensity of the radioactive emissions from different compounds of uranium depended solely on the amount of uranium present and therefore the phenomenon was a property of the uranium atom and not of the various molecules, such as the uranium oxides, which made up the different compounds. In searching for other radioactive

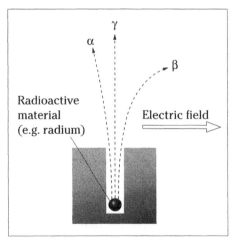

Radioactivity was investigated by burying some radioactive material in a shielding lead block and studying the rays it emitted. When subjected to crossed electric and magnetic fields, the alpha rays were deviated slightly in one direction the beta rays were deviated strongly in the opposite direction, and the gamma rays were not deviated at all. Subsequent studies showed that the alpha rays are positively charged particles identical to the nuclei of helium atoms, the beta rays are negatively charged electrons, and the gamma rays are ultra high energy electromagnetic photons.

elements, they discovered polonium – named after Madame Curie's country of birth – and radium, an extremely radioactive element which was to
become very important as a tool to probe the structure of the atom. Subsequent studies were to show that the radioactive emissions were of three
different kinds: alpha, beta and gamma rays. The alpha rays were bent
slightly by a transverse magnetic field, the beta rays were bent much more
but in the opposite direction, and the gamma rays were not deflected at
all. We now know that the first two are particles, the alpha particles being
the same as the positively charged nuclei of helium atoms, and the beta
particles being the lighter, negatively charged electrons that are the car-
riers of electrical current. These particles will be discussed further in the
next chapter, but the point to be made now is that the gamma rays are
electromagnetic waves of a much higher frequency and therefore a much
shorter wavelength and much higher energy than even the X-rays. Hence,
in the space of a decade, the known electromagnetic spectrum had been
extended immensely and now stretched from radio waves to gamma rays,
passing through the intermediate regions of infrared, visible, ultraviolet
and X-rays. The tremendous scale of this range can be demonstrated by
the changes in electromagnetic frequency that occur in passing through
each of the spectral regions. Frequencies in the infrared are a million times
higher than in the radio, the visible is a hundred times more than the infrared, the ultraviolet ten times more than the visible, the X-rays a hundred
times more than the ultraviolet, and the gamma rays another hundred
times more than the X-rays. The total range of frequency from radio to
gamma rays is a staggering 10 000 000 million. This compares with the spectral range then available to observational astronomy, which was restricted
to the visible and the neighbouring near infrared and near ultraviolet
regions, which covered a total range of frequency of only a factor of two
– an octave. But in the great astronomical period following the Second
World War, that factor of two was to be extended to the 10 000 000 million
quoted above, with the result that astronomy entered a new golden age
that we are still enjoying today.

At this point, the very end of the nineteenth century, it seemed that the
natural philosophy initiated by Newton, to be called classical physics, had
almost completed its task. There were two forces of nature, gravitation
and electromagnetism, whose properties were fully understood, and the
different forms of energy, particularly heat and light, had been completely
explained in elegant theories. There were some remaining problems, such
as the invariance of the velocity of light and the slight movement of the
perihelion of Mercury, but these were so subtle that it was confidently

119

expected that some small adjustment to current thinking would offer an explanation. Other non-subtle problems were posed by the fact that atoms emitted light at fixed discrete wavelengths, and that some elements spontaneously emitted strange energetic rays in the process of radioactivity, but these were regarded as mysterious properties of the indestructible atom, which would be revealed in time in the pursuit of classical physics. There was therefore a feeling that physics was nearing its ultimate goal of understanding the forces and energy of nature, at which point it could become an archive, a library from which any information about any physical phenomena could be extracted as required. Nothing could have been further from the truth; physics was on the verge of an explosion as deep and far-reaching as that caused by Galileo and Newton, so much so that natural philosophy up to this point was to be called classical physics and the new natural philosophy of the twentieth century was to be called modern physics.

CHAPTER SEVEN

The New Natural Philosophy

Darest thou now, Oh soul,
Walk out with me toward the unknown region,
Where neither ground is for the feet
nor any path to follow
Walt Whitman

Historically, natural philosophy meant the study of nature in all its forms, but following its great expansion in the nineteenth century, it was broken down into different disciplines such as astronomy, physics, chemistry and biology, which were themselves to be further subdivided. Natural philosophy tended to become a term associated with physics, and some universities, notably in Scotland, retained that name in preference to physics. But in the early part of the twentieth century, a major revolution was to occur in the subject which was to result in the renaming of the earlier developments, based on the natural philosophy of Newton, as classical (or Newtonian) physics, and the new natural philosophy of the twentieth century as modern physics.

Modern physics is a child of the twentieth century; indeed, it seemed to be waiting for the year 1900 in order to burst out of classical physics. It was to make extraordinary advances in the understanding of matter and energy, and at the same time to create two new basic theories of physics – quantum mechanics and relativity – which, to the general public, have unfortunately become two classic examples of the incomprehensible. The development of modern physics was of tremendous importance to the advancement of the understanding of astronomical phenomena, but of much wider importance were its implications for science and technology in general, because these were to lead to developments that were to have effects on the human way of life which were immense and far reaching. But there is also a great inherent beauty in its perceptions, in its theories and in its models, which leads me to claim that it represents the greatest cultural, as well as the greatest intellectual and scientific, achievement of the first half of the twentieth century.

At the end of the nineteenth century, there were three clues as to the inadequacy of classical physics, one of which was rather subtle and two others which were assigned to the mysterious nature of the atom. They have all been referred to in Chapter 6, but will be briefly summarized again here. The first of the atomic clues lay in the emission of light by atoms, which, when excited in a gaseous form, emitted light only at fixed frequencies and wavelengths unique to that atom. Studies of the much simpler problem of light emission from hot solid bodies, when the atoms and electromagnetic radiation are in equilibrium, marked the start of an epic series of experiments and theories that resulted in totally new concepts and which culminated with an understanding of the nature and structure of atoms in the wave formulation of quantum mechanics. The subtle clue was given by the Michelson–Morley experiment, which showed that the velocity of light was constant and independent of the motion of the observer. This led to the special theory of relativity, which in turn led to the general theory of relativity, both involving new revolutionary concepts. The third clue lay in the phenomenon of radioactivity, the strange energetic rays emitted spontaneously by elements such as radium (found in pitchblende), and the use of these radioactive rays was to lead to the development of nuclear physics, the study of the structure of the nuclei of atoms. All of these developments were of major importance to astronomy, to which they provided a great stimulus, as will be revealed.

The paper marking the start of the new physics, which was also the pioneering paper in quantum mechanics, was published in 1900 by Max Planck, the great German scientist after whom the prestigious German research institutes are named. He was born in Kiel and educated at the University of Munich, after which he went to Berlin University, where he was fortunate to conduct research under the famous physicists Helmholtz and Kirchoff. He was later to succeed the latter in Berlin in 1889, where he conducted the great researches that led to his award of the Nobel prize in 1918. He lived through both of the world wars of the twentieth century and had become famous enough to oppose the Nazi regime in the 1930s openly and without penalty. However, his son Erwin was to be executed in 1945 after being accused of participating in the plot to assassinate Hitler.

At the end of the nineteenth century there was no scientific basis for tackling the problem of the emission of electromagnetic radiation by atoms that occurred at several fixed frequencies and which were unique to that atom. The answer had to wait until the structure of the atom was revealed, but Planck took the first step by solving the simpler problem of the emission of light by solid bodies in which the light is trapped to an extent that it must come into equilibrium with the atoms before it escapes from the surface. The distribution of the velocities and energies of atoms in equilibrium with themselves had been determined with the kinetic theory of classical physics, which calculated when the entropy of the system reached a maximum as required by the second law of thermodynamics. In the case of a hot body, the light is also in equilibrium with the atoms of the body, its entropy must also maximize, and since the nature of light was fully understood by Maxwell's theory of electromagnetic oscillations, the condition of maximum entropy could be determined in order to calculate the distribution of frequencies emitted by hot bodies. The result showed agreement with observations at long wavelengths, say the red and infrared for a white-hot body, but deviated markedly at shorter wavelengths in the blue and ultraviolet.

The observations were on a very firm foundation. It was known that the radiation emitted by a hot body peaked at a certain wavelength and decreased on each side of that wavelength. At the temperature of the surface of the Sun, a body will emit most strongly in the green and the intensity will drop in the red and infrared on one side of the peak and in the blue and ultraviolet on the other side. If the temperature is increased, the level of maximum intensity shifts to shorter wavelengths and, if it decreases, the maximum shifts to longer wavelengths. Hence, a body heated from room temperature will first appear as dull red, then become yellowish and then white; if it were possible for the material to withstand even further increases in temperature, it would ultimately become blue like the hottest stars. In order to establish a condition when the material and radiation are in full equilibrium, the material must be able to absorb radiation at all wavelengths, in other words to be "black", but since matter can also emit at any wavelengths it can absorb, it will become white when heated sufficiently. This explains the somewhat confusing term applied to such radiation as "blackbody radiation", which is generally accepted and will be used subsequently in this book. Of course, to find a material that is totally black is not easy, but this problem was solved elegantly by constructing a spherical cavity that could be raised to any required temperature (below the melting point of the material). The radiation from the walls was completely trapped and would therefore come into full equilibrium with the material of the walls; to observe it only required a pinhole to be punched into the chamber and the ensuing radiation to be measured. The nature of blackbody radiation, in terms of its distribution with wavelength (colour) and its variation with temperature was therefore fully established. But it did not agree with that predicted by taking the well established electromagnetic theory of light and applying the well established second law of thermodynamics. Something was seriously wrong with one of these.

Planck proposed a solution in his seminal paper of 1900, which was remarkably simple in its expression but also remarkably difficult and revolutionary in its concept. He explained the observed blackbody radiation and its variation with temperature by assuming that light had the properties of a particle, as well as a wave: it could only exist in discrete quantities, that is that it was quantized. By making the energy of each quantum or wave packet proportional to the frequency of the light (or inversely proportional to its wavelength), he derived the constant of proportionality, which gave a perfect fit to the observations. That constant now bears his name and it joins Newton's constant of gravitation, Boltzmann's constant of thermodynamics and the velocity of light, as one of the fundamental constants of nature.

Planck was the forerunner of quantum mechanics, a theory generally regarded as incomprehensible to the wider public, but its name has a very simple origin: it derives from his proposition that light, despite its wave nature, acts in discrete quantities or quanta. This means that light has a dual character, acting both as an electromagnetic wave and as a discrete particle, or wave packet, which Planck called a quantum but which is now more generally referred to as a "photon". Despite Planck's precise derivation of blackbody radiation, his theory was not accepted in all quarters, because of the conceptual difficulty of assigning both wave and particle characteristics to electromagnetic radiation, characteristics that seemed to be totally separate in normal experience. Indeed, it is possible to argue that wave phenomena are impossible if treated from a quantum point of view. For example, the interference fringes produced by a light source behind two slits could easily be explained by a wave motion; the light was emitted in phase but, after passing through the different slits, the waves spread out (diffraction) and on meeting again their path difference caused phase shifts that resulted in bright fringes when in phase, and dark fringes when in opposition. But to explain this phenomenon in terms of light being composed of individual quanta or photons seemed quite impossible; a photon could go through one slit or the other and could not have any phase relationship with a photon passing through the other slit. There are other examples that express this dilemma in the inverse direction, where phenomena, such as the emission of hot bodies treated by Planck, can be explained only by regarding light as being composed of discrete entities and not of waves. But it has to be understood that light has *both* wave and particle characteristics, and can demonstrate either in any situation. The problem lies with us because it is a property of nature that is contrary to our own everyday macroscopic experience, and it was the first of many similar surprises that were to emerge as scientists probed into the microscopic world of matter.

Planck's seminal paper was to be followed by rapid and remarkable developments by many brilliant physicists, which were to establish quantum mechanics as one of the basic theories of nature and were to lead to an understanding of one of the most mysterious and basic agents of nature – the atom. The route is strewn with Nobel prizes and the story will be resumed later, but it is more appropriate at this point to relate the development of the other great new theory of physics – relativity – which spanned only a decade and, unlike quantum mechanics and atomic physics, was the creation of one individual, Albert Einstein.

At the same time that Planck was considering the introduction of the quantum to explain one inadequacy of the classical theory of light (the emission of hot bodies), another was being generated by two American physicists, Michelson and Morley, who, as mentioned previously, applied an interferometer of unprecedented accuracy to measure the change in the velocity of light caused by the motion of the Earth, whose velocity of 30 kilometres per second (about 20 miles per second) is 10 000 times less than that of light. In 1897 they published their final result, which confirmed an earlier experiment; the conclusion was that, whatever the direction of their light beam – with, at right angles or contrary to the Earth's motion – the velocity of light remained unchanged. This was in clear contradiction to the electromagnetic theory of light, which predicted such a change, since the Earth's motion simulated a wind through the ether, the medium that transmitted the light waves. The proposal by Planck that light had a particle as well as a wave motion did not help this problem because, if viewed as particles, the velocity of light would vary as the velocity between source and observer varied, but this had been shown not to be the case in the nineteenth century by astronomical observations of binary stars. The early studies of stellar spectra had revealed that many stars were binary in nature, that is, they were composed of two stars revolving about each other in gravitational union. This was revealed by the Doppler shifts in the spectral lines of either star as it orbited the other one, first approaching the Earth and then receding. When approaching, the light quanta would have an increased velocity and, when receding, a decreased velocity, and this would therefore destroy any phase relation between the two and cause them to reach the Earth at different times. No such loss of phase was observed; all binary stars showed a steady, periodic oscillation in velocity. Hence, the dual wave–particle nature of light did not help to solve the problem; if a wave, the result of the Michelson–Morley experiment could not be explained and, if a particle, the observations of binary stars could not be explained. On the basis of classical physics and, indeed, on the basis of common sense and experience, the constancy of the velocity of light, no matter what the motion of either the source or the observer, seemed impossible. The solution to this paradox

125

was to come from an equally paradoxical source – the Swiss Patent Office in Berne, where a young clerk was employed on the important but mundane tasks of patent checking. His name was Albert Einstein.

Einstein was born in 1879 in Ulm in the southern part of the newly formed united German empire. After schooling in Munich, he applied in 1895 for university entrance into the prestigious Swiss Federal Institute of Technology but failed the entrance examination. He then spent a year at a special school in order to obtain a diploma regarded as an entrance qualification by the Swiss Federal Institute, which he was then able to enter in 1896. His performance was good but far from outstanding and, on graduating, he found it difficult to secure the kind of appointment he wanted, which was a research position in a university, his many applications being either rejected or ignored, and for two years his only means of support depended on temporary teaching duties and an allowance from his parents. Abandoning the hope of a university position, he applied for and accepted an appointment as a technical expert (third class) in the Swiss Patents Office in Berne. This post was in an area and at a level consistent with his performance at university, and any hope of a research career would seem to have vanished. Yet, by the time he was 40, he had created entirely new concepts that linked space with time, matter with energy and gravitation with space, making him the first person ever to be regarded as being in the same genius class as Newton. Like Newton, he worked on his own; like Newton, his early achievements were made in his midtwenties in an intellectually infertile environment. But, unlike Newton, he was not a withdrawn character, always enjoyed the company of others, and published his theories as he made them without having to be persuaded to do so by some perceptive Halley.

In 1905, a year of immense personal productivity, Einstein published five papers, all of great importance but two of which were in the epoch-making category, since they established the special theory of relativity. This was based on two simply stated principles, the first of which was influenced by the ideas of the German scientist–philosopher Ernst Mach, who argued that absolute motion had no meaning and that the only meaningful motion was the *relative* motion between two bodies. This led Einstein to his principle of *relativity*, which states that the laws of nature are identical in all inertial systems, that is systems in uniform rectilinear motion with respect to each other without any external forces; by Newton's first law of motion, such systems move at uniform velocity in a straight line. Einstein's second principle derived directly from the Michelson–Morley experiment and states that the velocity of light is constant and independent of motion. As already stated, this invariance of the velocity of light was the subject of considerable scientific debate, since, on normal logical argument, it seemed to be impossible, to the extent that some scientists considered that the experiment was flawed. Einstein's approach to this paradox was more direct: if the velocity of light was *observed* to be invariant, then the *argument* that it was impossible had to be flawed and, since that argument was based on generally accepted experience, it had to be flawed in some very basic way; it was. Einstein was to show that a fundamental assumption made by Newton and all previous giants of philosophy – an assumption that was never stated to be an assumption because it was so self-evident and so ingrained in the human mind as to be unquestionable – was in fact incorrect: time was not absolute. Einstein did not reach this conclusion easily, and his attempts to apply his relativistic principles to the motion of inertial systems always foundered on the same rock – the constancy of the velocity of light. He was later to explain his great intellectual breakthrough with the words:

My solution was really for the very concept of time, that is, that time is not absolutely defined but that there is an inseparable connection between time and . . . velocity. With this conception, the foregoing extraordinary difficulty (the constancy of the velocity of light) could be thoroughly solved. Five weeks after my recognition of this, the present theory of [special] relativity was completed.

Relieved of the constraint of absolute time, Einstein was able to calculate how it changed as viewed by an observer in another body which was moving relative to it; there was also a consequent change in spatial dimension, which he also calculated. The direction of the changes was always the same; time was dilated, that is, it went more slowly in the other body, whether it was moving towards or away from the observer, and the magnitude depended only on the relative speed; similarly, space was always contracted, independent of the direction of motion and its magnitude depended only on the relative speed. These concepts were very difficult to grasp at that time (and probably still are for the lay reader) because the kind of velocities for such effects to be detectable lie far beyond human experience. For the Concorde, which flies at a speed in excess of sound, the effect is quite negligible and it would take 10 000 years of constant flying of Concorde at maximum speed to build up a time difference of one second. Even at the speed of the Earth of 30 kilometres per second, it would take nearly 70 years to build up a time delay of one second. Only at ultra-high velocities that reach a significant fraction of the velocity of light do the effects become significant. As a numerical example, and one again to illustrate the nature of the effect, let us take two space ships, in each of which we place an identical standard one metre rod and an identical accurate clock, after a comparison has shown the rods to be the same length and the clocks at the same time. Each is projected into space with an attendant astronaut and accelerated until they have a relative velocity of 87 per cent of the speed of light, a value selected simply because it gives dilation and contraction factors of two. If the two clocks are synchronized at 12 noon, say, then after one hour the observer in the first spacecraft will see a rod of one metre length and his clock will read 1.00pm But if he then observes the second spacecraft with an appropriate telescope, he will see a rod of half a metre length and a clock which reads 12.30pm Time has slowed and dimensions have shortened in the second spacecraft as seen from the first, but as far as the observer in the second spacecraft is concerned, his metre rod is still a metre and an hour is still an hour on his clock. These quantities are entirely relative and the situation is exactly reversible, so that if observations are made from the second system, then the first becomes the "moving" one, and it will see the identical contraction and time dilation in the other. If the velocity reached the velocity of light, the observed clock in the other spacecraft would stand still and time would be frozen; this answered a question that Einstein used to ask himself in his youth: "What would the world look like if I rode on a beam of light?".

Einstein then went on to examine the consequences of his theory of relativity for other quantities observed in a moving system, including the Doppler effect, which is of great importance to astronomy because it allows the velocity of celestial objects in the line of sight (the radial velocity) to be measured. In classical theory, if a spectral line is shifted to longer wavelengths by 1 per cent of its rest wavelength, then its velocity is in the direction away from the observer and its magnitude is 1 per cent of the velocity of light. This relation continues for 10 per cent, 20 per cent and more until, when the shift is 100 per cent, that is when the observed wavelength is double its rest value, the velocity deduced is the velocity of light. Although this classical theory is a very good approximation at low velocities, it deviates markedly from the relativistic relation at high velocities. Einstein's true derivation of the Doppler effect showed that, when the shift became 100 per cent (the wavelength had doubled), the velocity of the object was only 60 per cent of the velocity of light rather than equal to it, and when the shift became 200 per cent (the wavelength had trebled) the velocity was 80 per cent of the velocity of light and not twice as large (as the classical theory predicted). As the velocity of recession of any object approached the velocity of light, the observed wavelength would tend to become infinitely long and the frequency of the light would tend towards zero; from Planck's condition, this means that the energy of the light quantum approaches zero, so that any objects moving away at the speed of light would be undetectable. A related conclusion came from Einstein's evaluation of how the relative velocity appeared to two systems moving at different speeds relative to some reference. If two bodies (trains, aircraft or spacecraft) are set off in opposite directions, the classical theory says that the relative velocity between them is the sum of their individual velocities as seen from the starting point. But this is an approximation which is closely true only at low velocities. The relative velocity as seen from each system is always less than the sum of their individual velocities; it is also less than the velocity of light and can *never* exceed it. As an example, let us take the situation when two bodies are set off in opposite directions at half the velocity of light; then classical theory says that their relative velocity would be equal to that of light, but the true relative velocity would only be 80 per cent of the velocity of light. The maximum possible velocity between the two bodies is the velocity of light, at which point, as related above, communication between the two is no longer possible because the light no longer has any energy.

Einstein's first paper on special relativity was published in June 1905 and his later application of this to simple dynamics led to a major revelation which he hurriedly rushed into print in September of the same year. In considering the relative mass of a body in a moving system, he found that its increase with velocity was equal to the kinetic energy it was carrying (because of its motion), divided by the square of the velocity of light. A more rigorous consideration of the relativity of mass and energy led him to derive the general relationship between them, which has become the best-known equation in the world:

$$E = mc^2$$

where E is energy, m is mass and c is the velocity of light. Energy has mass, and mass is a form of concentrated energy. The laws of conservation of mass and energy are therefore manifestations of a wider law which says that it is their sum, expressed in mass or in energy, that is conserved. The two laws had been regarded as separate because the velocity of light is so large that the mass of the energy in the forms dealt with by the start of the twentieth century were so trivial as to be undetectable. This was true of the chemical energy involved in the burning of fossil fuels such as coal and oil, but with the development of nuclear energy (to be treated later in this chapter) the energy involved became large enough to be "weighed".

The consequences of special relativity were of great importance for science in general and astronomy in particular. It is instructive to remember that it is based on very simple principles, but principles that avoided the natural bias placed on the human mind by everyday experience. A psychological difficulty in accepting it was experienced by many, including scientists, and this was ultimately overcome by the large body of evidence that built up in its favour. This will be revealed as this story proceeds, but it is appropriate here to give two of the most direct examples. The most difficult concept to accept is that of time dilation, but this phenomenon

129

was to be clearly demonstrated by studies of cosmic rays, the extremely energetic atomic nuclei and electrons that pervade interstellar space and bombard the Earth. On striking the upper atmosphere, they collide with and smash many of the atoms there to produce "showers" of subatomic particles which are detected at ground level. Some of these particles are unstable and they decay into another form in a known time interval; they are little clocks. But many particles that are detected have such ultra-short life times that they should have decayed well before reaching the ground, even travelling close to the velocity of light. The effect was fully consistent with special relativity; their clocks had been slowed down considerably by their high velocity; their decay times had lengthened (relative to us), allowing them to reach the ground before decaying. The second example was to be provided by the development of man-made particle accelerators. In these, the increase in particle mass with velocity (because of its increased kinetic energy) could be measured and was in agreement with Einstein's $E = mc^2$. In many accelerators, particularly those involving electrons, the rest mass (i.e. the mass when stationary) of a particle is very small compared to the mass of its kinetic energy when fully accelerated to maximum velocity.

Despite the great success with his theory of special relativity and the impact he made on science in 1905, Einstein did not immediately acquire the university research post that was his main ambition, but the Swiss Patent Office in Berne did promote him to Technical Expert (second class) in 1906! Not until 1909 did he secure a university appointment as associate professor in the University of Zurich, whence he moved to Prague in 1911 as a full professor and then on to the University of Berlin in 1914. During this period, from Berne to Berlin, he started on his greatest work, the generalization of his theory of relativity, which was to include accelerations in his moving systems; this inevitably took him into gravitation and in 1915 he published his *general theory of relativity,* a new theory of gravity which, remarkably, was to replace Newton's. It was remarkable because Newton's theory, with its great initial achievement of explaining Kepler's laws of the motion of the planets, had then gone from strength to strength, from the explanation of the tides and the orbits of comets to a famous triumph in the nineteenth century when perturbations in the orbit of Uranus led to the prediction and discovery of a new planet – Neptune. It seemed unassailable; yet in 1919, a critical observational test came down in favour of the new theory of general relativity. The historical background to this story deserves mention, because the new theory was generated in Germany in the early stages of the most destructive war ever waged up to that time, but its confirmation was made by experiments organized by scientists from one of Germany's prime opponents – Great Britain.

In embarking on his general theory of relativity, Einstein's thoughts were greatly influenced by the property of gravity that was demonstrated by Galileo, a property that greatly intrigued him: all bodies, whatever their nature, whatever their mass, fell equally in a gravitational field. This had led Newton to make inertial mass (the mass with which a body resisted any changes in its motion) equal to gravitational mass (the mass with which a body responds to the force of gravity). Einstein believed that there must be some fundamental reason why these two masses were so identically equal and concluded that this must be because they were the same thing. He proposed his *principle of equivalence*, which equates accelerating systems with those in a gravitational field, thereby making his attempt to generalize the theory of relativity the same as a study of gravitation. The principle of equivalence states simply that a system in a gravitational field would be equivalent and indistinguishable from a system being uniformly accelerated at the same rate as a body would fall in that gravitational field, and that they would obey the same physical laws. Of course, if a human being were in a spacecraft sitting in a gravitational field or in a spacecraft being equivalently accelerated, it would be very easy to determine which, by simply looking out of the window, but in all other aspects, it would be the same.

The equivalence principle could therefore be used to study gravity by equating it to a uniformly accelerated system. In the case of gravity, a body that is dropped is pulled to the ground, such as Newton's apple, by the force of gravity. In the equivalent case of an accelerating system, a body that is "dropped" no longer experiences the acceleration being applied to the system, and it therefore remains "stationary" until the accelerating floor strikes it, and it is then carried forwards again as if it had a weight like in a gravitational field. The principle of equivalence states that these two situations are equivalent and all phenomena within them are identical.

An accelerating system can therefore be used to determine the equivalent gravitational effects, as Einstein was to do. If you could imagine that the room that you are in was not in a gravitational field but in equivalent uniform acceleration, then a projectile launched from one wall would appear to fall before striking the opposite wall, because the room is accelerating during its transit; but the projectile would consider itself to be travelling in a straight line! Also, if a beam of light is shone from one wall to the other, it would appear to bend because of the room's motion during its brief time of passage; hence, light should also bend in a gravitational field, a prediction of which effect was to become the most crucial test of the theory of general relativity. Another experiment to be conducted in

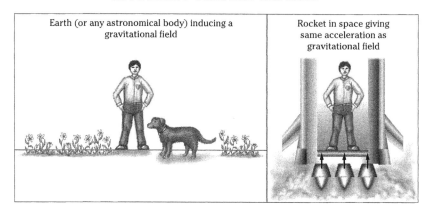

| Earth (or any astronomical body) inducing a gravitational field | Rocket in space giving same acceleration as gravitational field |

Any system in a gravitational field is equivalent to one in space, which is being accelerated at the same rate as the local acceleration of gravity. All laws and effects will be identical in the two systems, which are then indistinguishable (unless one looks out of a window!).

your room is simply to drop something (perhaps an apple), then the apple will no longer experience the force exerted by the accelerating engine (until it struck the floor); it would be force-free. The equivalent situation of a body dropped in a gravitational field meant that it would not experience any force; it would be completely weightless. These two factors – that a projectile passing through the accelerating room would consider it was travelling in a straight line, and any article that was dropped would not experience any force – led Einstein to a concept of gravity totally different from that of Newton. He proposed that the gravitational effect of mass was to distort the space around it and that bodies would move in *geodetics*, that is, they would follow the shortest distance in the curved space, in the same way that aircraft follow the shortest distance between two airports on the curved Earth. In the absence of, or at great distances from, any matter, space is no longer curved but flat, and bodies would then move in a straight line; this is Newton's first law of motion and Einstein's approach was to generalize that law for curved space.

The concepts developed by Newton and Einstein are so totally different that you would be fully justified in asking the question, "Which is the right one: is there a force of gravity or is space curved by the presence of matter?". The answer is a philosophical one: each approach is an attempt by the human mind to create a model to represent nature, gravity in this case, and which one is adopted depends entirely on the scientific test as to which one best explains and predicts what is observed. From what has been revealed so far in this story, either model would give the identical result for any gravitational effect, for example the motion of the planets, and any choice would have inevitably continued to lie with Newton, with his easier concept and simpler mathematics. It was only later in the development of his theory that Einstein made another breakthrough in understanding the nature of time – as he had done in his theory of special relativity – that a difference emerged.

Einstein conducted a thought experiment in which he used the principle of equivalence and his special theory of relativity. Three clocks are placed in an inertial system (one of which being subject to no external forces) and these are carefully synchronized. The first clock is fixed at a position above the second clock, and the third clock, to be used as a test clock, is placed at the same position as the first clock. The system is now subjected to a uniform acceleration and, at an appropriate time, the test clock is released from its support so that it no longer experiences the accelerating force and is effectively in free-fall. When it reaches the lower clock (or, if you prefer the equally valid point of view, when the lower clock reaches it) there will be a relative velocity between the two and the test clock will see a slower time in the lower clock as given by special relativity. From the principle of equivalence, this means that time goes more slowly in a gravitational field. Einstein was able to calculate that the time dilation in a gravitational field was the same as that given by special relativity for a body moving at the same relative speed as the velocity needed to escape from that gravitational field (the gravitational escape velocity). For the values of gravity known at that time, the effects were quite negligible; for example, in the case of the Earth, a time difference of one second would take fifty years to build up, compared to outer space where gravity was effectively absent. But at high values of gravity, not experienced then, the effects could become very appreciative and even dominant. Einstein's formula showed that, in the case of an object with a gravitational field so intense that a body on its surface could only escape if it had the velocity of light, clocks would stop and time would be frozen. Receipt of information from such an object would not be possible, its surface would represent an *event horizon* beyond which communication would not be possible; it was to be descriptively called a *black hole*. The possibility of such objects existing seemed totally remote; for the Sun to become a black hole, it would have to be contracted to a radius of three kilometres (about two miles) and the Earth to one centimetre (the size of a sugar lump). Although these two particular examples will never occur in nature, there is evidence that more massive systems could be in the black hole domain, as will be revealed later in this book when the major astronomical discoveries of the post Second World War years are described.

Having developed the concept of his theory of general relativity, Einstein, like Newton, needed an appropriate and sufficiently powerful mathematical technique in order to apply it; but, unlike Newton, he did not have to create it himself. The geometry of flat or straight space, which had been set up by Euclid in the time of ancient Greece, was quite inadequate for Einstein's needs, and he was very fortunate that the mathematics of curved (non-Euclidean) geometry had been developed, as a purely academic exercise, by some nineteenth-century mathematicians, of whom Riemann was the most notable.

With his new theory, now enshrined in appropriate mathematics, Einstein was able to tackle specific problems in gravitation in order to compare his theory with the Newtonian solutions to the same problems. His conclusion that time proceeded at a rate that depended on the gravitational field, being slower the higher the field, meant that he had to extend his earlier concept of matter distorting space to include time. His theory allowed the calculation of the trajectories of bodies (or light) in a gravitational field as geodetics (the shortest paths) in curved space and time; this involved four dimensions, three in space and one in time. Applying this to the orbit of Mercury, Einstein was immediately able to explain an effect that had baffled astronomers for 50 years, the advancement of its perihelion, which had been discovered by the French astronomer Leverrier and which had so puzzled him, as described in the preceding chapter. In Newtonian gravitation, the perihelion (the nearest approach of a planet to the Sun in its elliptical orbit) is fixed. In general relativity, a planet at perihelion experiences its highest gravitational field because it is at its closest approach to the Sun and its time will therefore be slower than in any other part of its orbit. This has the effect of reducing its velocity, which pulls its orbit slightly inside a Newtonian one, causing its perihelion to advance in the direction of its motion. Einstein calculated an advance of 45 seconds of arc per century in agreement, within the uncertainties, with the observed value. This was the first of what are now known as the three classical tests of general relativity.

The second was the prediction that light escaping from a gravitational field would be redshifted because the slowdown in its clock means a slowdown in the light frequency or an increase in wavelength, causing a redshift. However, this prediction was not a critical one, because it could also be made on Newtonian grounds. Light is a form of energy and, hence, from Einstein's mass–energy relation, it also has mass; therefore, it must expend energy in escaping the gravitational field, and therefore, from Planck's quantum criterion, its frequency must decrease and its wavelength increase. The predicted magnitude of the redshift is identical in each case. In any case, in astronomy, the gravitational redshift could not be disentangled from Doppler shifts caused by local velocities and it was not until the years following the Second World War that instruments of sufficient sensitivity were available to measure and confirm the effect in the laboratory.

The third test was the most crucial because it was capable of experimental verification at that time and, unlike the motion of the perihelion of Mercury, it was a prediction of a new effect and not an explanation of an old one. It has already been shown that light passing through a uniformly accelerated system would appear to bend and, as a consequence of the principle of equivalence, this required that light would be bent by a gravitational field. But the deduction of the bending of light could also be reached by Newtonian arguments because photons have energy; therefore, by Einstein's special theory of relativity, they have mass and their motion will be influenced by a gravitational field. But the magnitude of any deflection of a light beam in a gravitational field is predicted to be twice as large by Einstein than by Newton. In general relativity, the path of a light beam is a geodesic in space *and* time; when the gravitational field is highest, its clock is slowest, giving it a lower velocity which causes a bending equal to that which would be caused by the curvature of space alone and therefore doubling it. The Newtonian calculation did not allow for a variation in time and therefore gave a value half of that predicted by general relativity. The test laboratory was astronomical and was provided by the occasional occurrence of a total solar eclipse, when the Moon (which happens to have the same apparent diameter as the Sun when viewed from Earth) blocks out the Sun's light, thereby enabling stars in its direction to be seen which would otherwise be swamped. Such a solar eclipse would allow the positions of those stars that appeared close to the edge of the Sun, and whose light would therefore experience the maximum bending, to be measured by photographing them through a telescope. These positions could then be compared with the positions of the same stars when the Sun was not in the way and could not therefore induce any bending; this could be done at another time of year when the stars were visible at night. Any bending would, when observed at eclipse, reveal itself as a shift in the position of each star in a direction away from the Sun compared to its position outside of eclipse. Einstein's predicted shift was just under two seconds of arc (1.74, to be precise) whereas the Newtonian value was exactly half of that. On that tiny difference, less than one second of arc, rested the choice between two theories of one of the basic forces of nature – gravity.

The first eclipse expedition to conduct this experiment was a British one and was planned in the aftermath of the greatest war that their country, or anyone else's, had ever been engaged in. Its central figures were Sir Arthur Eddington, a leading theoretical astrophysicist, and Sir Frank Dyson, the then Astronomer Royal. The eclipse occurred over South America on 29 May 1919 and two expeditions were made to two separate locations in its path. Both had clear weather and successfully conducted the planned observations. The subsequent analysis showed shifts in the positions of stars near the solar limb, which were in the right direction (away from the limb) and whose magnitude was in agreement with Einstein's prediction of nearly two seconds of arc, and not the Newtonian value of half that amount. The announcement of this result caused a sensation, not only in scientific circles, but also with the public. Einstein's name became a household word and he reached a superstar status that no other scientist was to emulate.

Although of academic interest only, it is nevertheless within the spirit of this book, which seeks to present science as a human activity, to place this incident of light-bending in its historical framework. Einstein first concluded that light would bend in a gravitational field from consideration of his correspondence principle, which showed that a straight beam of light passing through an accelerating system would appear to be curved because of the motion of the system during its passage. Hence, a gravitational field would also cause a curvature, and a simple calculation for the case of the Sun gave a value just less than one second of arc, a prediction published in 1911. At that time Einstein must not have realized that the identical value could be derived by Newtonian arguments, but must have believed that the bending of light by a gravitational field was a result of his theory alone. Subsequent considerations, which have been outlined above, led him to realize that time slowed down in a gravitational field and that this would affect the degree of light-bending by the Sun. When closest to the Sun, time slows down slightly and this effectively reduces the velocity of the light beam, thereby causing an additional shift because of a process (analogous to the bending of light by a lens) in the material of which the velocity of light is reduced. Another simple calculation shows that this effect is in the same direction, and of an identical magnitude to, that caused by the curvature of space or of Newton's force of attraction. The fully developed theory of general relativity therefore predicted a shift of starlight by the Sun that was exactly twice that first predicted by Einstein in 1911. The point of interest is that two eclipse expeditions were mounted to test that 1911 prediction, one in 1912 and the other in 1914. The first, an Argentinian one, was to an eclipse over Brazil and was completely rained out; the second, a German one, was to an eclipse over the Crimea, but had to be abandoned because of the imminence of the First World War. Had either of these expeditions been successful, they would have produced a result twice as large as predicted by Einstein, and the subsequent advancement of his theory to include the variability of time would have appeared as a modification imposed by the observations and would not have had the same dramatic effect as that caused by the 1919 eclipse results. There is a great difference between the confidence generated by a theory that predicts a new effect correctly and one that explains an effect that is already known, as was the case in explaining the advancement of the perihelion of Mercury.

Tests of general relativity have been carried out with ever-increasing accuracy to this day, and all have been confirmatory. But one important prediction, that gravitational perturbations will produce a new form of radiation – gravitational waves – which propagate at the velocity of light, still eludes direct detection, although there is now some strong indirect evidence which will be presented in a later chapter.

Armed with his new theory, Einstein had the confidence to attempt, for the very first time, to model the Universe at large. At that time, the Universe was believed to be large but finite (meeting Olbers' paradox) and to be stationary (as evidenced by the fixed pattern of the stars over centuries). However, the theory of general relativity could not give a stable solution; the stars in a static Universe would inevitably collapse on one another and coalesce in one large mass. Einstein did not need his new theory to tell him this, because it had been clearly demonstrated by Newton; the Moon does not fall onto the Earth, nor does the Earth fall into the Sun, because they are not static, they are in motion. Any system of bodies will reach an equilibrium when the energy of their motions balances the force of gravity; if there is no motion (as in Einstein's cosmological model), the matter will collapse. Because the idea of a static Universe was so dominant, Einstein did not pursue a possible dynamic solution, but instead introduced his *cosmological constant* (an unfortunate term), which essentially meant the existence of another force, repulsive in nature, which would balance the attractive force of gravity at large distances. The cosmological constant has been debated at length since Einstein's paper of 1916, but the repulsive force it represents has never been detected and Einstein himself abandoned it some 20 years later for a reason that will be divulged towards the end of this chapter. However, in recent years, the concept of such a force has been postulated as existing in the very first instant of the start of the Universe, and this will be dealt with in the last chapter on modern cosmology.

General relativity was a great triumph but, unlike its great sister theory of modern physics, quantum mechanics, it had no great immediate impact on science. In astronomy it had explained two small effects, the advancement of the perihelion of Mercury and the bending of starlight by the Sun. But as astronomy developed in the first half of the twentieth century, except for cosmological models, it was Newton's theory that was used, because it was accurate enough for most purposes. This situation was to change dramatically in the post-war years, when the application of new technologies opened up the whole electromagnetic spectrum to astronomical observations and resulted in new and exotic objects being discovered which are fuelled by immense sources of gravitational energy. The general theory of relativity was to come into its own.

In parallel with Einstein's development of his theories of relativity, great progress was made in the study of quantum mechanics and the nature of the atom. At the time that Planck was developing his quantum theory of light, J. J. Thompson in Cambridge (England) was studying the rays produced in a cathode-ray tube. Earlier studies of the spectra of atoms had been made in discharge tubes into which gases could be sealed and an electric current passed through in order to make the gas fluoresce. The cathode-ray tube operates on the same principle, but in a vacuum at high voltage when rays are detected to emanate from the cathode (a white-hot heated filament) and cause fluorescence on a screen at the other end of the tube. By the simple and ingenious method of passing these rays through crossed electric and magnetic fields, he was able to measure the mass and the charge of the propagating particles. He obtained a mass that was very light, about 2000 times less than the mass of the lightest atom, hydrogen, and he obtained a charge that was negative and whose magnitude was to become the unit of atomic electric charge. It should be realized that the fact that it was negative was the result of convention only; with the development of the electric battery, electricity flowed from one terminal to the other and one was labelled positive and the other negative. This resulted in the particle that had been discovered and measured by J. J. Thompson, which was to be called the electron, being assigned a negative charge. He was later to detect positive rays in his tube, travelling in the opposite direction to the electron rays, and analysis revealed a particle mass that was much heavier and about equal to that of the hydrogen atom. He had discovered the proton. Its charge was positive but its magnitude was exactly equal to that of the electron; the charges of the two particles were opposite and equal.

These experiments raised the question as to whether the impregnable atom had a structure after all. The answer came in 1911 from Ernest Rutherford who was to develop the first ever model of the structure of atoms and then go on to initiate and develop the new science of nuclear physics; he was a giant in an age of giants. Born near Nelson in New Zealand in 1871, he took a degree at Canterbury College, which won him a scholarship to Cambridge, where he worked under the great J. J. Thompson. From there he went to McGill University in Montreal and then, in 1907, he was appointed to the Chair of Physics in the University of Manchester. It was there that he carried out the first breakthrough experiments in atomic and nuclear physics. Rutherford used the property of radioactivity to develop a powerful tool for his experiments. Placing a radioactive material, such as the element radium (discovered by Marie and Pierre Curie) in the centre of a dense shielding block, such as lead, which had a narrow tube drilled to its centre, constrained the radioactive rays to emit in a narrow beam. He could then separate the three types of emission, called alpha, beta and gamma rays, by the application of a transverse magnetic field. It was the alpha rays he was to employ to such singular effect and he was to show that these particles had four times the mass of J. J. Thompson's proton and twice the charge (they were later to be shown to be the same as the nuclei of helium atoms). Rutherford used the alpha particles as projectiles to bombard and penetrate atoms. He directed them into a thin gold film and obtained a surprising result: the bulk of them went straight through the film with hardly any deviation and a few recoiled almost exactly backwards. From this he deduced his model of the atom of being a tiny, positively charged, central nucleus containing nearly all of the mass and surrounded by a cloud of orbiting, negatively charged electrons. The atom was almost empty! But why had it appeared to be such an impenetrable structure in the past? The answer lay in the extremely high velocity of the electrons, which, like an aircraft propeller, present an effectively solid barrier to anything but the fastest moving microscopic bodies such as alpha particles.

The Rutherford atom was like a miniature planetary system. Like gravity, the force between the central nucleus and the orbiting electrons varied as the inverse square of their separation, and was attractive because the charges were opposite in sign (when electrical charges are of like sign, they repel, again as an inverse square law). Of course, a gravitational force also exists between the nucleus and electrons because of their mass, but this is totally negligible compared to the electrical force. So, it would seem that Newton's theory could be applied to the Rutherford atom, with a simple change for the scale of the force, and the orbits of the electrons would accordingly be generally ellipses or, in the special case, circles. But this was not so because the electromagnetic theory of classical physics had one major difference compared to the gravitational analogy of planetary motions. Unlike the planets, the acceleration of the orbiting electrons in the electric field of the nucleus would cause them to emit electromagnetic radiation at an intensity and wavelength that could be calculated exactly; this would cause them to lose energy and consequently to spiral inwards until they eventually coalesced with the nucleus. This clearly did not happen and it therefore marked some major deficiency in classical physics when applied to the microscopic world of the atom. This problem intrigued a young Danish theoretical physicist, who made a revolutionary proposal in 1913, while he was working with Rutherford in Manchester and only two years after Rutherford had revealed the electrical nature of atoms. The young man extended Planck's idea of the quantum nature of light to a quantum concept of the atom, and since the importance of quantum theory lies in its ability to describe the structure and properties of atoms, he is rightly regarded as the father of quantum mechanics. His name was Niels Bohr and on returning to Denmark he was to exercise a tremendous international influence over the development of atomic and nuclear physics.

Bohr approached the problem by considering the simplest possible atom, hydrogen, which consists of a proton around which a lighter, single electron orbits. This could be fully and easily treated by classical physics; from any starting position, the velocity of the electron and the shape of its orbit could be calculated precisely and this resulted in a decaying orbit (ending with a coalescence of electron and proton) and the emission of light over a continuum of wavelengths. Experiments showed that there was no coalescence, the atom was stable, the electron and proton remained apart, and the emitted radiation was at fixed discrete wavelengths whose spectrum, in the case of hydrogen, was in the form of an orderly series of converging frequencies. Bohr proposed a quantized model of the hydrogen atom with the following two properties: first, of the infinite number of orbits that classical physics expects an electron to have in revolving about a proton, only a limited number are actually allowed and these are circular and do not emit electromagnetic radiation (i.e. they are stationary and do not decay); secondly, the electron can jump from one stationary orbit to any other, thereby emitting (or absorbing if the second orbit has a higher energy) a quantum of radiation whose energy is equal to the energy difference between the two states.

Bohr now had to search for the criterion that would define his stationary orbits and he found it to be the simple one that the total orbital angular momentum of the electron had to be quantized in terms of Planck's constant; that is, the electron mass multiplied by its velocity multiplied by the circumference of its orbit had to be equal to Planck's constant or twice Planck's constant, or three times Planck's constant, or four times, five times and so on – intermediate values were not allowed. On this basis, Bohr was able to calculate the orbits and energies of all the states of the hydrogen atom and thereby derive its spectrum, which was in superb agreement with what was observed. The precision of the agreement, plus the fact that the calculation used values of the electron mass, electron charge and Planck's constant, which had been derived entirely independently, left no doubt as to the validity of the theory. An added confirmation was provided by the spectra of the hottest stars, which were known to contain a strong spectral line that could not be identified because it had not been seen in the laboratory. The Bohr theory provided an immediate identification; it was caused by the transition between the third and fourth orbits of the electron in ionized helium, that is, helium with one electron detached (because of the high temperature) which makes it hydrogen-like, but with a double charge in its nucleus, and easily subject to a Bohr

140

calculation. The corresponding transition between the third and fourth orbits in hydrogen itself lies in the infrared.

The Bohr quantum theory was quickly modified and extended by Arnold Sommerfeld, who showed that elliptical orbits, as well as circular orbits, were allowed for electrons in atoms, but these were also restricted by a quantum criterion that constrained the ratio of the sizes of the major to minor axes of the ellipse to be in integral numbers. This required another quantum number, which will be called the elliptical quantum number, as well as the principal quantum number defined by Bohr. In the lowest orbit of the hydrogen atom, when the principal quantum number is one, there is only one circular orbit, because the elliptical quantum number is zero. In the next higher state, when the principal quantum number is two, the elliptical quantum number is one and, in addition to a circular orbit, there is one elliptical orbit allowed whose major to minor axes are in the ratio of 3:2, being the sum of principal and elliptical quantum number over the principal quantum number. In the third quantum state, the principal quantum number is three, the elliptical is two, and there are one circular and two elliptical orbits, one of the same ratio as above and the other of 4:3 . . . and so on. Further studies showed that the electron and the proton had both a mechanical spin and a magnetic moment, each of which was quantized in direction, as well as magnitude, so that they could only lie parallel or anti-parallel to each other or to the corresponding effects caused by the motion of the electrons in their orbits. Hence, four quantum numbers were now needed to define the state of an electron in an atom.

An important new quantum principle was now introduced by Wolfgang Pauli, the Austrian–Swiss physicist, which has become known as the *Pauli exclusion principle* and which has stood up to every test it has been subjected to. It states that in any atom, only one electron can occupy any individual quantum state at any one time; all other electrons are excluded. This means that only one electron can have the same values of the four quantum numbers. With this exclusion principle and the Rutherford–Bohr–Sommerfeld atom, quantum mechanics was about to make the most profound contribution to chemistry that that subject has ever experienced. Until that time, in the early 1920s, chemistry had been a largely empirical discipline, but a considerable amount of order was put into it by the brilliant work of the Russian scientist, Dmitri Ivanovich Mendeleev, who proposed a periodic table of the elements in 1870. He did this by investigating their chemical properties in relation to their atomic weights, which could be measured chemically and which always came out to be a whole number if hydrogen is listed as one. He noted that, if he arranged the elements in order of their atomic weight, a periodicity developed in their chemical properties. He then assigned a successive number to his ordering of the elements, starting with one for hydrogen, and this number is known as the atomic number. Each element is therefore designated by two numbers, its atomic number and its atomic weight, the latter usually being about twice or more of the former. It was found that most elements had more than one atomic weight, even though they were chemically identical; that is, they had the same atomic number. Hence, hydrogen with atomic number 1 and atomic weight 1 also had a less abundant *isotope* of atomic weight 2, called deuterium and referred to as heavy hydrogen; carbon, with atomic number 6, had isotopes with atomic weight 12 and 13. In the first three rows of the periodic table (see p. 143) it is the most abundant isotope that is listed. The number above each element is the atomic number and the number below each element is its atomic weight. The elements in bold print have been discovered since Mendeleev's original table.

The elements in each column have similar chemical properties and, as Mendeleev proceeded to construct his periodic table, he found that he sometimes needed to leave a gap in order to maintain the periodicity and the flow of atomic weights. These he proposed to be due to undiscovered elements, whose properties and weights he could predict with considerable accuracy, thereby stimulating searches that led to their discovery in the laboratory. One of these was germanium, which has properties like carbon and silicon and was named after the country in which it was discovered. Then, in University College London, Sir William Ramsay discovered the inert gases helium, neon and argon in the closing stages of the nineteenth century, and these added another column to Mendeleev's table. As recounted earlier, helium had been discovered spectroscopically in the Sun some 30 years earlier by Sir Norman Lockyer, who proposed it to be a new element and named it after the Sun.

Mendeleev's periodic table became an accepted and central part of chemistry. It was now fully explained by the Bohr–Sommerfeld atom and the Pauli exclusion principle. The atomic number, the successive number given to the elements in progressing through the periodic table, was the number of electrons in the atom, which was determined by the number of protons in the nucleus in order to give a neutral combination. The periodicity of the table was explained by the number of electrons allowed to be present in each shell as defined by the principal quantum number. The first shell consists of a single circular orbit which only has two elementary states and which can therefore only contain two electrons, whose spins must be antiparallel. This shell becomes filled as soon as the atomic number reaches two (helium) and this causes it to be chemically inactive and inert. The second shell can carry two electrons in its circular orbit, but it also has an elliptical orbit, which has six elementary states, allowed by a combination of spins and angular momenta. This will also become filled at atomic number ten (neon), which is therefore also inert like helium.

The first part of the periodic table.

^{1}hydrogen$_{1}$							^{2}helium$_{4}$
^{3}lithium$_{7}$	^{4}beryllium$_{9}$	^{5}boron$_{11}$	^{6}carbon$_{12}$	^{7}nitrogen$_{14}$	^{8}oxygen$_{16}$	^{9}fluorine$_{19}$	^{10}neon$_{20}$
^{11}sodium$_{23}$	^{12}magnesium$_{24}$	^{13}aluminium$_{27}$	^{14}silicon$_{28}$	^{15}phosphorous$_{31}$	^{16}sulphur$_{32}$	^{17}chlorine$_{35}$	^{18}argon$_{40}$
			^{32}**germanium**$_{73}$				

The other non-inert elements between atomic numbers one and ten have "vacancies" in their shells, which allow them to combine with other atoms to form molecules. When such combinations fill a shell, the molecule is very stable. For example, oxygen has two vacancies left in its shell and can combine with two hydrogen atoms to form the stable molecule of water (H_2O), and carbon has four vacancies which allow it to combine with four hydrogen atoms to form another stable molecule, methane (CH_4). When the third shell is reached, there is again a circular orbit, but two elliptical orbits are also allowed. The circular orbit can carry its two electrons, the first elliptical orbit can carry its six electrons, and the second elliptical orbit could carry ten electrons. But these electrons in the second elliptical orbit require an energy higher than those in the circular orbit of the next shell, the fourth. Hence, the next electron added moves into the fourth shell and the third shell will act as if it is entirely filled when only its circular and first elliptical orbit are filled. This happens when a further eight electrons are added beyond neon, making the next inert element argon, with an atomic number of 18. Of course, the filling of the second elliptical orbit in the third shell would have to happen ultimately and this explained the greater complexity of the periodic table at high atomic numbers.

Quantum mechanics had developed a model of the atom which could explain both its physical and chemical properties and, although this was a tremendous achievement, there was unease about its artificiality. Why did nature select and allow only certain electron orbits and why did she choose criteria based entirely on whole numbers? The answer came quite quickly, in 1924, when the French theoretician Louis de Broglie proposed that particles have a wave nature, as well as a discrete character. Planck had taken the well established wave phenomenon of light and proposed that it also had the properties of a particle. De Broglie was proposing that the well established concept of matter, the particle, also had a wave character. The circle was complete; the old concept of matter and energy as completely separate entities was becoming blurred; Einstein had in any case shown the two were related by $E = mc^2$.

De Broglie proposed that the wavelength of any particle could be determined by equating the classical expression for its momentum (mass times velocity) with the quantum expression derived from Planck's postulate for the energy of a light quantum. This gave the wavelength to be Planck's constant divided by the classical momentum. Planck's constant is so small that for any material particles, even a grain of sand, the wavelength is essentially zero and it behaves in the way we conceive a particle to behave. It is only at the microscopic level, in the atom, that the quantum effects become appreciable and even dominant. That is why they escaped attention for so long. The behaviour of electrons in atoms is dominated by their wave nature and they can only exist when the circumference of their orbit is one wavelength, or two wavelengths, or three wavelengths and so on. Using the de Broglie wavelength in this way gives the Bohr condition that was developed for the hydrogen atom. The discrete orbits of the electrons and their quantum nature was therefore explained in terms of this new wave mechanics of quantum theory. The other particle characteristics of the quantum theory, such as elliptical orbits, can also be explained in terms of the electron's wave characteristics.

The full formulation of the new quantum mechanics was carried out by an Austrian, Erwin Schrödinger, who in 1925 published his famous wave equation, which was to be extended in 1928 by Paul Dirac, who included the effects that arose from the special theory of relativity. In that same year, the wave property of electrons was demonstrated directly by G. P. Thompson (son of J. J. Thompson) by passing a beam of electrons through a thin foil and observing fringes on the other side caused by the diffraction and interference of the electron waves. This allowed the wavelength of the electrons to be measured, and by varying the beam velocity and hence the energy and momentum of the electrons, it was shown to be fully consistent with de Broglie's expression.

Another consequence of the wave nature of particles was developed by the German physicist Werner Heisenberg, who pointed out that the position and momentum of a particle could not be determined with exact precision; an uncertainty of a fundamental kind existed in these two quantities which was imposed by the wave nature of matter. In his *uncertainty principle*, he quantified this effect in stating that the product of the uncertainty in the position of a particle and the uncertainty in its momentum had to be, at best, equal to Planck's constant. Because of the very small value of that constant, the uncertainties are completely negligible for normal material particles; for example, the position and momentum of a grain of sand can be determined precisely, but the situation for an electron is quite different. In the first Bohr orbit of the hydrogen atom, the electron completes a full single wavelength and it cannot be said to be in any one part of that orbit at any one time; its uncertainty in position is its whole orbit! Heisenberg's uncertainty principle is one of the more difficult concepts to grasp, but it is a direct result of the wave nature of matter and the theory of quantum mechanics. This theory was to have the most profound consequences for the development of astrophysics and it provided the basis for studies that would ultimately allow both the physical conditions and chemical composition of celestial bodies to be determined from their spectra.

In completing this story of the development of the wave theory of quantum mechanics, I cannot resist thinking back 25 centuries to the ancient Greek mathematician and philosopher, Pythagoras (see Ch. 3), who established the link between whole numbers and the harmonious

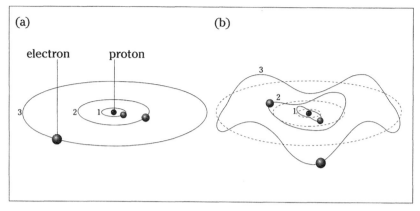

The hydrogen atom consists of a positively charged proton about which orbits the much lighter, negatively charged electron. The fact that it only emits (or absorbs) specific discrete wavelengths of light was explained by Bohr by the fact that only specific, discrete electron orbits were allowed and these were defined by whole numbers. (a) Shows the first three orbits defined by numbers 1, 2 and 3. The explanation came from de Broglie who proposed that electrons had a wave as well as a particle nature (it had already been shown by Planck that light had a particle as well as a wave nature). (b) In the microscopic world of the atom, the electrons were acting as waves and the allowed orbits were those whose circumferences were equal to 1, 2, 3, etc. electron wavelengths, thereby allowing standing waves to occur. The emission or absorption of light photons is caused by a jump from one orbit to another.

notes produced by standing waves in taut strings. He believed that this link would extend to the whole of nature and, looking outwards, proposed that the motions of the planets could be explained by the fact that the spheres carrying them (as was then thought) would, like his strings, be harmonious. He was wrong, but if he had been able to look inwards, into the microscopic world, he would have seen his harmony, in all its glory, in the structure of the atom.

☆　☆　☆

You will remember that the great breakthrough in understanding atomic structure came in 1911, when experiments using alpha rays from radium showed that atoms consisted of a tiny, positively charged nucleus surrounded by a cloud of negatively charged electrons. This led to the developments already described and it culminated in the major triumph of the wave-mechanical quantum theory. The experiments were conducted by Ernest Rutherford, a New Zealander born in 1871, who did nearly all of his work in England, first in Manchester and then in Cambridge. He had that deep insight into natural phenomena and also the ability to reach the right conclusions from simple experiments, which marked him as the greatest experimental physicist since Michael Faraday. He was to attract about him a great team of experimental physicists who were to establish a new subject: nuclear physics.

Much of Rutherford's work was carried out by doctoral students (times have not changed so much), to whom he pointed the way with unerring foresight, left them to get on with it and caught up at thesis time. After the 1911 experiments, while still in Manchester, he continued using the alpha particles (helium nuclei) from radium as projectiles to probe the atom and gave a young doctoral student a project to study the penetration of these rays in air, a seemingly uninspired suggestion by the great man. The student set about his task, which needed only a radioactive source, a phosphor screen and a dark room. The alpha particles (with energies up to a million electron volts: 1 MeV) would strike the screen and cause a scintillation which could be recognized by a dark-adapted eye. He would then measure the number of scintillations as he varied the distance of screen from source. This was a somewhat tedious and soporific exercise and one day, feeling a little sleepy, he pulled the screen out of range of the rays and dozed off. On opening his eyes, still fully dark-adapted, he thought he saw another scintillation, discarded this as imagination and dozed off again, but, on opening his eyes a second time, he again saw a scintillation. He was now fully awake and concentrated on the screen, and he started to count the scintillations, which were at a much slower rate than when close up. All his previous data told him that the alpha particles could not reach the screen, so the question "What am I seeing?" became the whole purpose of his thesis. He concluded from these measurements that the unknown particles were hydrogen nuclei (protons), struck head on by the alpha particles and propelled farther to reach the screen; such hydrogen atoms would be available in the water vapour molecules present in the air. His data were consistent with this picture, so he wrote his thesis and was awarded his PhD.

Rutherford then took over in 1917 with a majestically brilliant series of experiments reminiscent of those conducted by Faraday, almost a century before, which established electromagnetism. He put the radioactive source that emitted the beam of alpha particles, together with a screen to detect them, into a sealed glass tube which he could fill with selected gases. On introducing hydrogen he found that the distant scintillations did not occur; his student's conclusion was wrong. Nor was there an effect when he introduced oxygen and, of all the possible constituents in the air, it was only when he introduced nitrogen that he detected the distant scintillations. Then, following J. J. Thompson, his mentor when he was a PhD student, Rutherford subjected the unknown particles to orthogonal magnetic and electric fields and, from the magnitude and direction of the induced deflection, he was able to measure their charge and mass. They were high-energy protons! A discharge section was then introduced into the tube so that the spectrum of the contained gas could be monitored. The experiment was then repeated with the introduction of nitrogen gas into the tube and the triggering of the radioactive source. At the start only the spectrum of nitrogen was seen, as expected, but after a sufficiently long time Rutherford saw the spectral lines of hydrogen and oxygen appearing, along with those of nitrogen, and their intensity continued to increase with time. The alpha particles were breaking into the nitrogen nuclei, which were then transforming into oxygen with the emission of a high-energy proton, the particle first detected by Rutherford's student. The actual interaction can be written as:

^{2}Helium$_4$ (the alpha particle) combines with ^{7}nitrogen$_{14}$ to form ^{8}oxygen$_{17}$ plus ^{1}hydrogen$_1$ (the proton)

Here, the first upper number is the atomic number, representing the charge of the nucleus, and the second lower number represents the atomic weight. The sum of each of these has to be the same before and after the reaction because the total charge and total mass are conserved.

Rutherford's great achievement was that he had split the unsplittable indestructible atom; he had transmuted the elements, the dream of alchemists over centuries, but in a way that they, including Newton, could never have imagined. A new discipline, nuclear physics, had been born. The series of experiments that led to this seminal discovery were triggered by a chance and wholly unexpected observation, not by Rutherford, but one of his research students. This reminds me of a pertinent comment, made many years before in the nineteenth century, by one of the great pioneers of biological research, Louis Pasteur, who remarked that "in the field of observation, chance favours only the mind that is prepared".

The year was 1919 and, shortly afterwards, Rutherford was persuaded to move from Manchester to Cambridge, where he succeeded J. J. Thompson as director of the Cavendish Laboratory. There, he built up and attracted a team of gifted scientists and pursued a programme that greatly extended his studies by the bombardment of various elements with alpha particles. Like his bombardment of nitrogen, these usually resulted in a nuclear transformation that involved the emission of a proton. But when the very light elements of beryllium and boron were tried, another unknown particle was emitted which had the property of great penetration. This was investigated by James Chadwick, one of Rutherford's staff, who was able to show that it was an electrically neutral particle with no charge but with a mass almost identical to the proton. It was a new particle, to be called the neutron (for obvious reasons), and Chadwick's discovery of it was to earn him a Nobel prize. For boron, the particular interaction that revealed the neutron was

^{2}helium$_4$ combines with ^{5}boron$_{11}$ to form ^{7}nitrogen$_{14}$ plus ^{0}neutron$_1$

You will note again that charge and mass are conserved in the reaction.

The neutron was discovered in 1932 and it then seemed that the basic structure and properties of all matter could be explained as a combination of only three elementary particles: the electron, the proton and the neutron. What simplicity! Each atom had a positively charged nucleus composed of protons and neutrons in a strong bond, and around it orbited negatively charged electrons whose number was equal to the number of protons, thereby making neutral the atom as a whole. In each element, this number (of protons or electrons) is the atomic number which locates the element in the periodic table and explains its chemical properties as the quantum theory demonstrated. The second number that defines each element is its atomic weight and this is given by the sum of the number of protons and neutrons in its nucleus, and since these have equal mass (or very nearly so) this explains why the atomic weights of the elements are always whole numbers. To take the two simplest examples, hydrogen has one proton as its nucleus, giving it an atomic number of one and an atomic weight of one (^{1}hydrogen$_1$), helium has a nucleus composed of two protons and two neutrons, giving it an atomic number of two and an atomic weight of four (^{2}helium$_4$). A similar assignment of the number of protons and the number of neutrons in the nuclei of all elements can be made from their atomic numbers and atomic weights. With the exception of hydrogen, with its single proton as a nucleus, all nuclei of the elements are formed by protons *and* neutrons, clearly demonstrating that the neutron is essential for binding the different particles together in each nucleus.

Another puzzle of the periodic table, the isotopes, also fell into place at this stage. Most elements exhibit more than one atomic weight and these are called isotopes, because they have identical chemical properties. For example, 99 per cent of carbon has an atomic weight of 12 (^{6}carbon$_{12}$) but 1 per cent has an atomic weight of 13 (^{6}carbon$_{13}$). This means that there are two stable nuclei of carbon, one with six protons and six neutrons and the other with six protons and seven neutrons, but since the number of protons is the same in each case, the number of orbiting electrons will be the same and, hence, the chemical properties will be the same. Another example is hydrogen which usually has a nucleus consisting of one proton but can also have a nucleus consisting of one proton plus one neutron. This isotope of hydrogen is called deuterium and, since it has twice the mass of ordinary hydrogen, it is also called heavy hydrogen.

A final subtle property of the periodic table also became understood because of these developments in nuclear physics. The atomic weights of the elements turned out to be whole numbers, for the reasons given above, but there were slight deviations from atom to atom which had defied explanation. The reason was that the energies needed to bind together the protons and neutrons in each atomic nucleus were different and therefore caused slight differences in mass because $E = mc^2$. The fact that the differences in mass were detectable meant that the energies in atomic nuclei were huge compared to all previous experience; they could actually be weighed! The nuclei of atoms are therefore bound together by enormous forces, which were completely unknown before Rutherford split the atom; its great strength explained why the atom had been considered indestructible in the past. A new force had been revealed, to be called the *strong nuclear force*, and it joined the gravitational and electromagnetic forces as a third force of nature (a fourth, the *weak nuclear force*, had been revealed earlier by the phenomenon of radioactivity). The new force was very different from the two established forces of gravitation and electromagnetism because, obeying an inverse square law, they were long range in their operation, but the strong nuclear force is extremely short range; even the electrons orbiting the nucleus of an atom do not experience it. In order to bind a nucleus, the nuclear force has to overcome the strong repulsive electric force between the protons and only does so at very short range. The simplest analogy is a highly compressed mechanical spring (which represents the repulsive electric force) which is held in a vice (the nuclear force).

The nuclear energy available in each atom could easily be calculated from its measured atomic weight and Einstein's equation, $E = mc^2$. This showed that the element with most energy (per unit mass) was hydrogen and, if this could be built up into helium and then into heavier elements such as carbon, nitrogen and oxygen, energy would continue to be released until the element iron (atomic number 26) was reached. Iron has the most tightly bound nucleus of all the elements and therefore has no accessible nuclear energy; lighter elements have more energy and heavier elements have more energy. So, if it were possible to transform the elements from one to another, energy would be available by building up the lighter ones towards iron (fusion) and by breaking down the heavier ones towards iron (fission). This variation in the nuclear energy of the elements was to become a critical factor in considering the birth, life and death of stars, as will be seen in a later chapter on stellar evolution.

The very simple picture that there were only three elementary particles in nature, the electron, proton and neutron, did not last very long. In the same year that saw Chadwick's discovery of the neutron in the laboratory, the American scientist Carl Anderson and British physicist Patrick Blackett independently discovered a new and different particle in the "showers" caused by the impact of the extremely energetic cosmic rays on the

Earth's atmosphere. It had a mass identical to the electron and a charge that was exactly equal but opposite. It was the anti-particle of the electron and is called the positron. If the two come into contact, which they will do readily because of their strong electrical attraction, they will annihilate each other and all of their mass will be transformed into pure energy, compared to the 1 per cent that occurs in the case of the fusion of hydrogen into helium. The discovery of the positron was the first clue to the fact that, in addition to the electron having its anti-particle, the proton and the neutron may also have their anti-particles. This has turned out to be the case, so in addition to the matter that is our everyday experience and of which everything, including ourselves, is made, there is an exactly opposite form of anti-matter which can be revealed by energetic particle collisions. The fact that our Universe chose to be made of matter rather than anti-matter is an important question in modern cosmology to which a convincing answer has not yet been given.

At about the same time as the neutron and the positron were discovered, another new and quite different particle was proposed to exist by Wolfgang Pauli with the same brilliant perception with which he developed his exclusion principle. He noticed that in many nuclear reactions, such as those caused by the bombardment of elements by alpha particles, there were small departures from conservation of momentum, leading him to conclude that another unseen particle must be ejected in the interaction and its "kick", or momentum, caused the apparent departure. Since the departures were very small, the particle had to be very low in mass; since the total charge in all reactions was conserved, it had to be electrically neutral; and since it could not be detected, it had to have a very weak interaction with matter. Pauli's hypothesis of the existence of a new, very light, neutral, weakly interacting particle had to wait 25 years before it was verified by its detection in 1956. It is so light that its mass has still eluded measurement to this day, although it has been established that it is at least as much as 10000 times less than that of the electron (which you might remember is 2000 times less than the lightest of all elements – hydrogen) The only force it experiences is the weak nuclear force and it is so weakly interacting that it can pass almost unhindered through the Earth and even the Sun. It is called a neutrino and it plays a very important role in astronomy, as the later chapters will reveal.

In the early 1930s, the probable number of elementary particles had grown from three to seven (the electron, the proton, the neutron, their antiparticles and the neutrino), but with the increase in the sophistication of instrumentation and the development of man-made accelerators, pioneered in the Cavendish Laboratory (Cambridge, England) by Cockcroft and Walton to produce high-energy particles in a more controlled way for the bombardment of atomic nuclei, a very large number of new particles were revealed. It was discovered that the proton and neutron, unlike the electron, were not elementary particles at all; they had structure and were composed of even smaller entities. Their study, to be called particle physics, continues until today and is one of the most fundamental areas of physics, since it still pursues the question as to the basic nature of matter. Its story can be stopped here in the mid-1930s, but it will be referred to again in the latter stages of this book when it becomes of great relevance to the big bang theory of the origin of the Universe.

The developments in nuclear physics, described above, were of major relevance to astronomy and will be returned to again in the discussion of stellar evolution and cosmology. Other developments occurred in the mid- to late 1930s, which were not astronomically relevant, nor were they more than interesting in a scientific sense, but their impact on world affairs, which continues today, was so immense that it is appropriate to relate the story here – the development of the atomic bomb. It starts with the question that was raised immediately the enormous scale of the energies locked inside the nuclei of atoms was revealed: could they be harnessed for human use? An unambiguous answer of no was given separately and independently by Bohr, Einstein and Rutherford, a remarkable trio of geniuses who, it was thought, could not possibly be wrong; but they were wrong, because of a property of one rare isotope which they could not have possibly foreseen. After Chadwick's discovery of the neutron, the great Italian physicist Enrico Fermi embarked on a series of nuclear experiments in Rome, which exploited its zero-charge property and which were to result in his award of a Nobel prize, which he received in Stockholm in 1938. He was accompanied by his Jewish wife, and the political climate in Italy at that time was such that he never returned to Rome, but emigrated to the USA. He was one of many gifted scientists who left their homeland to escape oppressive political regimes in Europe and, in so doing, greatly strengthened the scientific core of America.

Rutherford had made his great advances by bombarding atoms with high-energy alpha particles, and Fermi decided to do the same with neutrons, but since these neutral particles did not have to overcome the strong repulsive electric force that presented a barrier to the alpha particles, he slowed them down in the belief that this would make them more likely to stick to the nuclei. This exercise in common sense (an attribute of most great scientists) proved to be right and he readily added a neutron to the test atom, with the frequent result that it would disintegrate by shedding a small fragment, usually an alpha particle. This had the result of transmuting the target element to one of lower atomic number, but one of Fermi's aims was to try and make elements of higher atomic numbers, which would require the emission of an electron after the absorption of a neutron. Also, if he were to achieve this with uranium, it would mean the creation of an entirely new element to those existing on Earth or, to any-one's knowledge, anywhere in the Universe. To this end, in 1934 Fermi started to bombard the heaviest element, uranium (atomic number 92), with his slow neutrons and obtained a very surprising but puzzling result. An abnormal amount of energy was released, but its form could not be ascertained (there were no alpha particles) and the by-products of any nuclear transformation could not be determined. Although Fermi was cau-tious, the Italian press was not and proclaimed the creation of a new atom, the first transuranic element – it was wrong.

In Berlin, this result attracted the attention of a brilliant German phys-ical chemist, Otto Hahn, who was director of the Kaiser-Wilhelm Institute. Hahn had had a long interest in radioactive elements and their chemical properties, and had been fortunate in his apprenticeship years to work with Ramsay at University College London in 1904 and with Rutherford in 1905 during his period at McGill University in Montreal. Hahn repeated the Fermi experiment with uranium and, not surprisingly, obtained the same result of an abnormal energy release. But he had the chemical skill to con-duct a detailed chemical analysis of the byproducts and he found one new element, but only one, in addition to uranium: barium. This was a very surprising result because barium is well removed from uranium in the periodic table and its production would imply that the uranium atom was being split into two, whereas all previous nuclear transformations had involved small adjustments with the emission of a proton or alpha parti-cle. But if it was a nuclear transformation, as it had to be, where and what was the other bit or bits?

Hahn was assisted in his work by Lise Meitner, an Austrian physicist who was one of the few women, like Madame Curie, to break into the male dominated world of science. She was also a Jew but was protected, at that difficult time for Jews in Germany, by the fact that she was an Austrian cit-izen. But with the *anschluss* of Austria by Nazi Germany in 1938, she auto-matically became a German citizen and, no longer having the protection of Austrian citizenship, she fled to Sweden. There, with another German refugee, Otto Frisch, she considered the implications of Hahn's detection of barium and concluded that the uranium atom (atomic number 92) must be splitting into two large fragments, one of which was barium (atomic number 56) and, to conserve charge, the other fragment had to be krypton (atomic number 36) which, being inert, explained why Hahn had not detected it chemically. This large jump from uranium towards the most tightly bound of all atoms, iron, explained the very large energy release. It was also realized that this fission process only occurred in the rare lighter isotope of uranium (235), which was less than 1 per cent of natural uranium, the bulk being in the form of uranium 238. It was further realized that, since the total number of neutrons in the stable nuclei of barium and krypton was less than the number needed to bind the nucleus of uranium, about three neutrons must also be released in each reaction. The awesome prospect immediately dawned that the harnessing of nuclear energy, in the form of fission, was a possibility if a chain reaction could be arranged, that is, if the three neutrons produced by the fission of one uranium atom could split a further three atoms, producing nine neutrons, and so on, there would be such a multiplication that the energy would be released almost instantaneously. The result would be a bomb of unimaginable power. These findings were reported by Niels Bohr at an international conference held in the early part of the traumatic year of 1939. At that moment, nuclear physics left the world of basic science, the only one that it had known, and entered the world of high politics and, before long, of major warfare.

As you well know, these findings led to attempts to produce an atom bomb during the Second World War, one of which was fully successful. In 1942 the USA launched Project Manhattan under the nuclear physicist Robert Oppenheimer and this followed two possible routes. The first was to isolate the rare uranium 235 isotope, a particularly difficult task since it was chemically identical to the much more abundant 238 isotope, and could only be accomplished on a scale of almost atom by atom. If a sufficient quantity could be isolated and assembled, there was a critical mass (which had to be calculated) above which it would explode. This is because neutrons are lost at a rate proportional to the surface area, whereas energy is generated at a rate proportional to the volume; since the former scales as the square of the dimension and the latter as the cube, the latter will overtake at some point. So, if subcritical masses are assembled and then imploded to form a supercritical mass, an atom bomb results. The second approach had the same concept but searched for another fuel. Fermi's dream of building transuranic elements was realized in the early 1940s and the element beyond uranium with atomic number 93 was named neptunium after Neptune, the next planet beyond Uranus. Similarly, element 94 was named plutonium after the next planet, Pluto. Having run out of planets, the next transuranic elements were named in more ad hoc ways; for example, element 100 was named fermium after Fermi himself. It was found that plutonium had a fissile property similar to uranium 235 and was therefore another candidate for a bomb. It could be formed by the neutron bombardment of uranium 238 to produce plutonium 239 and this could be done in an atomic pile of natural uranium where the neutrons from the fission of uranium 235 would produce plutonium 239 by collision with the more abundant uranium 238. Both approaches were successful and on 6 August 1945 a uranium bomb was dropped on the Japanese city of Hiroshima; on 9 August, a plutonium bomb was dropped on Nagasaki; the result is now history.

These fearsome weapons ushered in an even higher realm of destructive power in the post-war years by using them as the detonator of the hydrogen bomb, in which they produced the ultra-high temperatures needed to fuse hydrogen into helium, as in the stars. Bombs have been made with the unbelievable force of being equivalent to 60 million tonnes (60 megatonnes) of TNT explosive making the thousand-bomber raids of the Second World War appear trivial in comparison. Fortunately, no nuclear weapons have been used in anger since Hiroshima and Nagasaki and there is a worldwide agreement that they never should be used. It is highly unfortunate that the possibility of such weapons was revealed by basic curiosity-driven research in nuclear physics, and it is salutary and remarkable to realize that it all stemmed from a single effect which was not of exceptional scientific importance or interest – the neutron-induced fission of the rare isotope of the heaviest naturally occurring element, uranium, and the consequent production of a few free neutrons.

The creation of the new natural philosophy described above, and now known as modern physics, was one of the greatest of human intellectual achievements. It was effectively established in just over 30 years and covered a period that saw the most destructive war ever fought on this planet, and it had powerful but unhappy inputs into a second, even more destructive, war. But it was also a time of intense scientific excitement and it involved many highly gifted individuals who picked up several Nobel prizes. Among them, the real giants were, in the order they first appear in this book, Planck, Einstein, Rutherford, Bohr, Pauli, de Broglie, Schrödinger and Heisenberg. It is almost impossible to make a further selection from a group of such exceptional talent, but if forced to, I would pick Einstein and Rutherford. Both became equally famous in the world of science but, for some reason, Einstein's fame extended to the general public, almost like a film star, with his photograph being familiar to everyone. Why this should be so is not easily explainable, but the reasons must lie beyond science itself, because Rutherford's scientific achievements were just as epoch-making, indeed they were more immediately relevant. Rutherford was the great experimentalist who opened up the microscopic world, he revealed the electrical structure of matter, he split the atom and transmuted the elements, he discovered a new force of nature, he created nuclear physics and prepared the way for particle physics, the study of the ultimate structure of matter. In contrast, Einstein was the great theoretician who looked at the macroscopic world in a more perceptive way, he established a link between space and time, and between matter and energy, he created a new theory of gravitation which displaced Newton's, and prepared the way for theoretical attacks on the ultimate problem of astronomy, cosmology, the study of the nature and evolution of the Universe.

CHAPTER EIGHT

Astronomy in the Early Twentieth Century

The riches of heaven's pavement, trodden gold
Than aught divine or holy else enjoyed
In vision beautific
Paradise Lost

The developments in nuclear physics in the mid-1930s, outlined in Chapter 7, provided a clue to one of the greatest puzzles in astronomy, one that had baffled astronomers and physicists for well over a century – the source of the Sun's energy. If the Sun was made of coal, it would burn itself out in about a thousand years, whereas historical records showed that it had hardly changed in ten thousand. This led Hermann von Helmoltz and William Thomson, the Professor of Natural Philosophy in Glasgow University, to propose in the nineteenth century that the source of the Sun's energy was gravitational in nature. Thomson was later to be made a peer and this is only mentioned here because he became better known by the name he then adopted as Lord Kelvin. Among other things, he established the absolute scale of temperature which bears his name (degrees Kelvin) and which is based on the centigrade scale but with a zero at the absolute zero of temperature (when atoms or molecules have no motion) rather than the freezing point of water which on the Kelvin scale lies at 273 degrees.[1] Kelvin's theory was that the Sun was still under gravitational contraction, thereby supplying the energy to keep it hot. His estimate of the Sun's lifetime was several hundred million years, far in excess of historical records, but geological sciences were developing rapidly at that time and, from fossil records, deduced an age of the Earth, and therefore the Sun, of thousands of million years. The geologists were right and it was only when the scale of nuclear energy was revealed in the 1930s that a likely explanation emerged. The largest single step in releasing nuclear

---

1. Temperatures expressed in this book are in degrees Kelvin unless otherwise stated.

energy was afforded by the fusion of hydrogen into helium, when 1 per cent of the matter would be released as pure energy, an immense amount. The Sun was known to be composed mainly of hydrogen (like the giant planets), it was very hot at its surface (about 6000°) and it was known that this meant that it was extremely hot in its interior. What if it was so hot that some of the nuclei of hydrogen (protons) had such large velocities that they could overcome the electrical repulsive forces and fuse together to ultimately build helium and thereby release the enormous energy of the nuclear force? Since the nucleus of helium consists of two protons and two neutrons, whereas four hydrogen nuclei give four protons, this required some transformation of two of those protons into neutrons, perhaps by the absorption of an electron. The actual processes are somewhat complex and had to wait some time for their elaboration, but the amount of energy that would be released by any such fusion process was large enough to be convincing; it would allow the Sun to shine for many thousands of million years.

This possible explanation of the enormous energy source of the Sun and stars had been recognized by leading astrophysicists in the 1920s, well before Chadwick's discovery of the neutron. The fact that four protons and two electrons, if they were able to combine by some process not understood then, would result in a helium nucleus 1 per cent less massive than the original constituents, indicated an energy source of the required scale. The actual fusion mechanism was not even remotely understood at that time, but it was known that any initial step would require protons to have velocities high enough to overcome the strong electrical repulsive force of another proton before it could combine with it; this could be achieved only if the temperature in stellar interiors were sufficiently high. The first models of the Sun and stars were produced in 1925 by the British astrophysicist Sir Arthur Eddington (of 1919 eclipse fame), whose work formed the basis for the study of stellar structure. He considered a star to be in equilibrium because the high gas pressure in its interior supported its weight; he also investigated the opacity presented by the star to the radiative energy released in its centre, before it was able to diffuse outwards and finally escape from its surface. This enabled him to calculate the distribution of density and temperature throughout the interiors of stars. For the Sun, he obtained a temperature in the core of about 10 million degrees Kelvin, very high but not high enough to give enough protons sufficient energy to scale the strong electrostatic barrier of another proton. Eddington concluded that the proposal was a tentative one. About a decade later, after critically relevant developments in quantum mechanics and nuclear physics, related earlier, a solution was provided in the mid- to late 1930s by the work of two brilliant theoretical physicists, the German-born Hans Bethe and the Russian-born George Gamow, both of whom, like many other scientists, had emigrated to the USA from a troubled Europe. The problem was how two positively charged protons could overcome the immense repulsive force at temperatures of 10 million degrees Kelvin in order to fuse; how could they cross the very high electrical barrier in order to reach the very deep attractive well of the short-range nuclear force and thereby release the enormous energy involved. One answer was to postulate even higher temperatures in the solar interior, but Eddington's models seemed to rule this out. The answer lay in the dual particle/wave nature of matter, as revealed by quantum mechanics in the form of the expression by de Broglie. As a particle, a proton would have to surmount the electrostatic barrier completely in order to cross it, and this was the basis of the earlier calculations. But, acting as a wave, it was able to tunnel through the barrier at levels below the peak height, allowing fusion to occur at lower velocities and therefore at lower temperatures. The work of Bethe and Gamow also outlined the detailed sequence of binary processes that ultimately resulted in the thermonuclear fusion of hydrogen into helium. The energy of the Sun and the stars, the energy of the Universe as then known, had been identified.

The inputs to astronomy caused by the development of modern physics in the first part of the twentieth century were of great importance, but in parallel with these there were major developments in astronomy which were the direct result of exploiting the observational power afforded by the spectrometer, the photographic plate and the several new, large reflecting telescopes of the twentieth century, which used glass mirrors coated with highly reflecting silver films. The largest of these was the 100-inch (2.5 metre) telescope on Mount Wilson, just outside Los Angeles, which was commissioned in 1917 and whose construction was the result of the efforts and vision of George Ellery Hale and the generosity of private American benefactors.

Most of the studies were concentrated on stars, the building blocks of the Universe. At the end of the nineteenth century, Annie Cannon had recorded the spectra of thousands of stars at Harvard College Observatory and she grouped these into several types which she placed in a sequence. Subsequent measurements of the colours of stars showed that Cannon's sequence was also a colour sequence, ranging from blue to yellow and then to red. Because of the pioneering analyses of stellar spectra by astrophysicists, prominent among whom were the Englishmen Arthur Eddington (already mentioned as one of the architects of the 1919 eclipse expeditions, which confirmed Einstein's prediction of the bending of starlight) and James Jeans, together with the American, Henry Norris Russell, it was shown that the sequence of stellar spectra and colour was attributable almost entirely to a variation in the temperature of the outer atmospheres of the stars. It was shown that the red stars had atmospheres with temperatures of about 3000° (Kelvin) and the very blue stars had atmospheres whose temperatures were about 50000°; the white–yellow Sun had an intermediate temperature of about 6000°. Another important development in these early stages of astrophysical analysis was made by Cecilia Payne of Harvard College Observatory. She was able to demonstrate that, although the stars were made of many different elements, there was one that was much more abundant than all the others – hydrogen.

The masses of different stars were also determined by the study of binary systems, in which two stars are bound gravitationally and revolve around their centre of gravity by Kepler's laws. Their velocities were measured by the Doppler effect in their spectra and the variation through a whole cycle revealed the period of revolution from which the circumference of the orbit and hence the size of the system could be determined. With velocity and size, Newton's theory was able to give the mass of the two stars and their spectra allowed these to be related to stars in general. These data showed that the range in mass varied from about one-tenth of a solar mass to about 50 times a solar mass and that the lightest stars were the coolest, the heaviest stars being the hottest.

The next important step was to determine the luminosities or total energy flux being emitted by the stars and this required a measurement of their distance, a critical parameter in astronomy but a difficult one to determine. The trigonometrical method of parallax, the angular change in position of a star, when the Earth moved in six months from one position to its opposite position around the Sun, allowed stellar distances to be measured out to about a hundred light years. But the observation of the spectra of the thousand or so stars within that range, whose distances and therefore luminosities were known, allowed the luminosity to be established for each spectral type. Hence, luminosities could be determined from their spectra for stars beyond the range of trigonometrical parallax, and their distances could then be determined from their apparent brightness. This method of determining distance, called spectroscopic parallax, greatly extended the distance range and was the first stage of a bootstrapping process, which was to be continually used until the ultimate scale of the Universe was reached. These spectroscopic parallaxes gave distances of only approximate accuracy, but they were enough to allow stellar luminosities (i.e. the total intrinsic energy flux from the stars) to be derived from their apparent brightness. The range of luminosities was very large: the most luminous stars were the blue ones and were emitting radiation at a rate of about 10000 times that of the Sun, and the least luminous stars were the faint red ones which were emitting at a rate 10000 times less than the Sun. There was a definite sequence in the properties of stars, running from the heavy, hot, luminous blue stars to the light, cool red stars, and, although the former were much more massive (up to about 50 solar masses) than the latter (down to about a tenth of a solar mass), their energy output was much greater than the ratio of masses involved. Hence, the mass ratio of 500 results in an energy output ratio of about a hundred million. When it was realized that the energy source of the stars is the thermonuclear burning of hydrogen into helium, which releases 1 per cent of the mass as energy, it was easy to calculate the total amount of energy available in a hydrogen star as one per cent of mc^2 using m as the mass of the star. From this, the longest possible lifetime of any star (when it had used up all of its hydrogen fuel) could be calculated. The results showed that the hot blue stars had lifetimes as short as millions of years; hence, the hot blue stars now seen, like the three that form the belt of Orion, were formed recently on astronomical timescales and even as recently as the appearance of the first human species on Earth. These early simple estimates were very approximate, but later estimates, based on a detailed consideration of stellar evolution, show that the lifetime of yellow stars such as the Sun is about 10000 million years, and of the faintest red stars, about 10000000 million years.

But in addition to the more numerous stars discussed above, there were other stars which did not fit into this sequence, notably the red giants and the white dwarfs (see Colour Plate 6). The former were emitting up to a million times more than their counterparts of the same colour and temperature lying on the main sequence, thereby indicating that they were extremely large cool stars, and the latter were emitting up to 10000 times less than their counterparts, indicating that they were extremely small hot stars. The red giants, of which a prominent example is Betelgeuse (the bright red star in the top left-hand shoulder of Orion) have sizes that exceed the orbit of the Earth, and the white dwarfs are as small as the Earth itself. The reason for this variation in the properties of stars was one of the very central problems in astronomy, since they form one of the main building blocks of the Universe, and it was to be largely solved in the post-war years when the whole life-history of stars became broadly understood. Stellar evolution will be discussed in Chapter 11, when it will be shown, among other things, that the Sun, a yellow star, will eventually expand into a red giant, engulfing the Earth in the process, and will ultimately contract into a white dwarf. But it will take 5000 million years for this process to start, so there is a long way to go.

Another area of intense astronomical study in the early stages of the twentieth century was the structure and particularly the size of the Milky Way, which was thought by many at that time to be the whole Universe. In the nineteenth century, the Milky Way was considered to have the shape of a flattened disk of the order of 100000 light years in diameter and about 5000–10000 light years thick. It was believed to constitute the entire Universe, beyond which there was a void and, as a large but finite system, it satisfied Olbers' paradox. The size had been estimated by very crude methods, such as assuming that all stars were as intrinsically bright as the Sun, but with the greatly improved distance-measuring techniques of the early twentieth century, a better attack on determining the size of the Milky Way became possible. This was a result of one of the really major steps in distance measurement, which was provided by the studies of Henrietta Leavitt of Harvard College Observatory. I cannot help commenting on the clearly male-dominated world of science at that time.[2] In this story so far, only five women have been mentioned and they did not appear until the very end of the nineteenth century and the start of the twentieth. Two, Marie Curie and Lise Meitner, made their contributions to radioactivity and nuclear fission respectively, and they worked at different institutes in different countries of Europe outside their own; the other three – Annie Cannon, Henrietta Leavitt and Cecilia Payne – made their contributions to stellar astronomy and they were not only in one country but also in one institute – Harvard College Observatory.

Henrietta Leavitt conducted a major study of variable stars and extended this to the Southern Hemisphere, where she observed the stars in the Magellanic Clouds, which were named after the Portuguese navigator, Ferdinand Magellan, who observed and recorded them in 1520 during his voyage around the world. There are two of these clouds, the Small and the Large, and they are small satellite galaxies to our own Milky Way, but that was not known at the time. However, it was realized that the stars grouped in each of them would lie at about the same distance from the Earth and therefore their relative luminosities would be the same as their relative brightness. This fact was exploited by Henrietta Leavitt in studying the variability of the stars within them. She found one kind of star whose brightness varied in a very regular fashion and the period of variation was closely linked to the brightness in a way in which the stars with the longest periods (about a hundred days) were about a hundred times brighter than those with the shortest periods (about one day). These stars are now known as cepheid variables, after the prototype which is the fourth listed star in the constellation of Cepheus (delta Cephei). As will be discussed when we reach stellar evolution, the cepheids are stars in a late

---

2. Some readers may think that this is still the situation today, although less so.

162

stage of their life when the gravitational control of their energy generation by thermonuclear burning becomes a little unstable. But their great importance to astronomy was their ability to determine distance by simply measuring the period of variation in brightness, which gave the luminosity, and then calculating their average apparent brightness; the inverse square law, by which the intensity of light decreases, did the rest. Also, since the total flux from the star could be used for the measurement (it did not have to be spread out in a spectrum) and since these stars turned out to have great intrinsic luminosities, they could be detected to very large distances and they afforded the greatest single step in extending the scale of astronomical distance measurement beyond those afforded by trigonometrical and spectroscopic parallaxes. The relationship is known as the cepheid period–luminosity law, and its value to astronomy will become apparent. There is one star in the sky which is a cepheid and which is known to every reader: Polaris, the north pole star, which is a yellow giant of the same colour and temperature as the Sun, but several hundred times more luminous.

But one important step was needed to turn the period–luminosity relationship into a tool for astronomical distance measurement. It had to be calibrated so that the relative luminosities could be turned into absolute luminosities and this could be achieved by determining the distance of any one cepheid, in the Magellanic Clouds or in the Milky Way. This problem was first tackled by Harlow Shapley, an American astronomer who became director of Harvard College Observatory. Since there is no cepheid close enough to measure its parallax and thereby determine its distance, Shapley had the brilliant idea of using the motion of the Sun as a baseline rather than that of the Earth. Relative to the surrounding stars, the Sun is moving at 20 kilometres per second in the direction of the constellation of Hercules, so in ten years it will have moved 20 times more than the diameter of the Earth's orbit, which is the baseline for the normal trigonometrical measurement of parallax. This would allow distances to be measured that were 20 times greater, but only if the stars were stationary, which they are not. Shapley overcame this difficulty by taking the average parallax of groups of stars which included a cepheid and which lay in a direction perpendicular to the Sun's motion; a greater distance had been reached but at the expense of accuracy. Nevertheless, Shapley's calibration, which showed that the cepheids were more luminous than the Sun by factors running from several hundred to several thousand, provided the most powerful tool then available for measuring the distance of very remote objects.

Shapley applied this to his great motivation, the determination of the size and structure of the Milky Way, which he believed, with many other astronomers, constituted the whole Universe. At that time, the Milky Way was believed to be made of stars in the form of a flattened disk, which was some thousands of light years in diameter and in which the Sun lay near the centre. But on either side of the Milky Way, and lying above it, were several clusters of stars, called globular clusters because of their spherical shape, and many of these included cepheid variables. Shapley studied these in both the Northern and Southern Hemispheres and, by using his newly calibrated period–luminosity law, he was able to determine their distance and distribution. He found that the globular clusters were not distributed evenly in the sky, that is, that the Sun was not located at their centre, which lay in the direction of the constellation of Sagittarius and was at a staggering 30000 light years away. He concluded that this represented the centre of the Milky Way system, and time was to show that he was right; the globular clusters surrounded the Milky Way in a roughly spherical distribution and the flattened Milky Way was as much as a 100000 light years across, with the Sun lying in an outer region, some two-thirds of the distance from the centre to the edge. In Copernican style, Shapley had removed the Sun from the centre of the Universe. The reason why it had been thought to lie at the centre was not to be revealed until 1930 when Robert J. Trumpler, a Swiss-American astronomer, showed that dust as well as gas was present in the plane of the Milky Way; this caused obscuration and restricted observations to a distance determined by the dust and not by the extent of the Milky Way. This meant that only the local part of the Milky Way was being seen and this turned out to be only the 20 per cent near the rim. Since the globular clusters lie above the plane of the Milky Way in which the dust is concentrated, they were not obscured and Shapley was able to observe them to their outermost limits.

Shapley's findings were published in 1917, and, soon after then, independent and quite different studies, by Jan Oort in Holland and Bertil Lindblad in Sweden, furnished supporting evidence. They were studying the motions of stars in the solar neighbourhood that had been measured by the Doppler shifts in their spectra, caused by their velocities in the line of sight to the Sun. They found that, in addition to the small random velocities of the stars, there was a systematic effect which could be explained as a circular motion at about 220 kilometres per second about a centre in the same direction and at about the same distance as claimed by Shapley for the centre of the Milky Way. At this point, in the very early 1920s, the broad structure of the Milky Way had been established and its scale had been determined with significant accuracy for the very first time. It was a disk consisting mainly of stars, but with an interstellar medium composed of gas and dust, of about 100000 light years across and a few thousand light years in thickness; it was rotating about its centre, causing the Sun, which lies about two-thirds of the distance from centre to rim, to have a velocity of 220 kilometres per second, and at that rate it will take nearly

300 million years to complete a full revolution; there was also a more sparsely populated spherical halo containing the globular clusters extending to many tens of thousands of light years above the plane of the Milky Way.

But was this vastly extensive system of the Milky Way the *whole* Universe, beyond which there was nothing? This, in the early 1920s, was one of the burning questions in astronomy, and the question could be framed in a more specific way: what is the nature of the spiral nebulae? The nebulae listed in the catalogues of Messier and the Herschels had been objects of great interest and study, and were generally recognized as lying within the Milky Way system; they were generally the consequence of stars in various stages of their life history, as will be described later in the chapter on stellar evolution, but the spiral nebulae were different. Whether they lay inside or outside the Milky Way system could not be established and, as you may easily imagine, a controversy arose between those who believed the former and those who believed the latter. This led to a now famous debate, held in the US National Academy of Sciences in 1920, between Harlow Shapley (who had established the scale of the Milky Way) and another American astronomer, Heber Curtis. Shapley argued for a location within the Milky Way and Curtis for one beyond the Milky Way. This debate revealed that the evidence available was quite insufficient to decide between the two possibilities, but the assembly of those facts in favour of its view by one side and the assembly of those facts in favour of the view of the other side, although justified by being highly entertaining and informative, does greatly question the value of such debates in reaching an evaluation of the truth. Returning to the particular issue of whether the spiral nebulae were inside or outside the Milky Way system, this could only be settled by one factor: determination of their distances. These were to be furnished by an American astronomer who was to join Hipparchus, Tycho Brahe and the Herschels as one of the greatest observational astronomers of all time. His name was Edwin Powell Hubble.

Hubble was born in Missouri in 1889 and had an unconventional early academic development, as has characterized many great scientists. He took a degree in physics at the University of Chicago, but then totally switched his field to take a degree in law at Oxford. But after a few months in that profession, he completely abandoned a career in law and returned to Chicago to take a doctorate in astronomy, which he obtained in 1917. He then spent two years in the US army before joining the staff of the Mount Wilson Observatory in 1919 at the age of 30, when most other scientists would already have made their mark. Of course, he was fortunate in gaining access to the recently commissioned 100-inch telescope, the largest and most effective in the world, but he was to exploit it with a brilliance that was to take astronomy beyond the Milky Way and towards the edge of the Universe.

Hubble used the 100-inch telescope to make extensive studies of the brightest spiral nebula, which lies in the constellation of Andromeda. He was able to resolve and observe its individual stars and he detected some with the very regular variation in brightness that marked the cepheids. He determined their periods and, from Shapley's calibration of Henrietta Leavitt's period–luminosity law, he was able to determine the distance to the great Andromeda nebula. He obtained one million light years (the latest estimate is 2.5 million) and this was sufficient to place Andromeda well beyond the confines of the Milky Way as determined by Shapley. The Milky Way was not the whole Universe; it was one of many island Universes like Andromeda. The scale of the Universe had once again increased by an enormous factor; the date was 1924 and extragalactic astronomy had been born.

The term nebula has all but disappeared from the description of extra-galactic systems and they are now referred to as galaxies. But the spiral galaxies, although the most numerous, are not the only kind of extragalac-tic system, as was demonstrated by Hubble, who classified them into three main types with subclassifications within them. In addition to the spiral galaxies there were also the irregular and elliptical galaxies, named, like the spirals, after their structural appearance (see Colour Plate 7) . The spirals comprise about 75 per cent of the external galaxies, the ellipticals comprise 20 per cent and the irregulars 5 per cent. The range of the total mass of the elliptical galaxies is very large, varying from 100 000 solar masses to 10 000 000 million; the masses of the spirals and irregulars fall within this range. Similarly, the range in size of the elliptical galaxies is much greater than that of the spirals and irregulars, varying from about a tenth of the diameter of the Milky Way (itself a spiral galaxy) to about ten times that diameter. There is therefore no pattern in the masses and sizes of the different types of galaxy, nor is there a pattern in their luminosities or star numbers. The only patterns that emerged lay in their colours and the amount of interstellar material they contained: the ellipticals are reddish and essentially devoid of an interstellar medium; the spirals are yellowish and contain about 10 per cent of interstellar gas; whereas the irregulars are bluish and contain about 25 per cent of interstellar material. The colours of the galaxies are the average of the colours of all the stars they contain, and this was to lead to an explanation of the link between colour and type. When stars are formed by gravitational contraction within the interstellar medium, a range of masses results and it is the heavy ones that become the brightest, the hottest and therefore the bluest, but their very high energy output also makes them the shortest-lived stars, lasting perhaps only ten million years. So, these hot blue stars will be present now only in galaxies that still have an interstellar medium to form them. Hence, they are more numerous in the irregular galaxies than in the spiral galaxies and are now absent in the ellipticals, thereby explaining the range in colour. The discerning reader will recognize that this means that all galaxies will tend to become redder with time as they use up their interstellar matter to form stars, but this does not mean that the irregulars will develop into spirals and the spirals into ellipticals; the relation between galaxy type and colour is not a simple evolutionary one, although collisions between galaxies, which are known to occur, may play a crucial role. Since it is believed that the galaxies were formed at about the same time in the evolution of the Universe, and since there is no relation between important initial parameters such as total mass, the reason for the variation in the nature and properties of galaxies is still a great puzzle today. Even the starting point in unravelling that puzzle has not yet been reached, the answer to the question "How are the galaxies formed?". This is one of the fundamental questions in modern cosmology and will be discussed in the final chapter of this book.

Hubble, who had made the gigantic step of demonstrating that the Universe was far more extensive than the Milky Way, which was just one galaxy among many, was to take an even greater step in a programme of observations that he completed in the early 1930s. He systematically observed the spectra of all the spiral galaxies within range of the 100-inch telescope and measured their radial velocities (the velocity in the line of sight) by means of the Doppler effect. To do this, he had to identify a spectral line whose wavelength he could compare with the same wavelength in a stationary source, data that had been well established by the laboratory physicist. Since the spiral nebulae were very faint, the quality of his spectra was rather poor and he was able to distinguish only one spectral line with any confidence, a line produced by ionized calcium (calcium with one electron detached because of the temperature of the stars in each galaxy), which was strong for a reason of atomic physics: it was produced by the lowest level in the atom (such "resonance" lines usually occur in the ultraviolet). Hubble found that the wavelength of the ionized calcium line was always shifted towards the red, that is, that the galaxies were receding from the Milky Way; further, he found that the velocity of recession increased as the galaxies became fainter and, since on average the fainter galaxies were more distant, he had established that the Universe was expanding. A galaxy that was a quarter as faint, and therefore (on average)

was twice as far away as another galaxy, was receding at a velocity that was twice as fast. Expansion is an effect like an elastic string being stretched apart or a balloon being blown up. The relative velocity between any two points, on the string or balloon, is proportional to the distance between them. Hubble had established the expansion of the Universe in a rough and relative way by assuming that the faintness of a galaxy is attributable entirely to its distance; this would be true only if all galaxies are equally luminous, but this is far from being the case. Hubble now had to tackle that fundamental recurring problem in astronomy – the determination of distance to an even greater scale than hitherto. He had measured the distance to the spiral galaxy in Andromeda by observing the cepheid variables it contained, but Andromeda was the nearest and brightest spiral galaxy, and most of the others were too remote and therefore too faint to allow cepheids to be detected and measured with the technology available at that time. Some other objects had to be used as distance indicators, objects that had to be very much brighter than the cepheids and whose luminosities had to be known: a tall order. The method adopted by Hubble and by other astronomers was a simple but very approximate one: the brightest star method. In studying stars in the Milky Way, where distances and therefore luminosities were reasonably well known, it was found that the hottest bluest stars were the most luminous by an amount up to a maximum of about 100000 Suns. So, in each galaxy, the brightest star was identified and assumed to have the same luminosity as the most luminous stars in the Milky Way. This "bootstrapping" method was to be extended later to use even more luminous objects, such as supernovae.

Hubble was able to determine the distances of about thirty galaxies out to ten million light years and, from their Doppler redshifts, was able to make the dramatic announcement in 1929 that the Universe was expanding. This was confirmed in the years following, when he and his colleague Milton Humason extended the range of expansion to nearly 100 million light years. We live in a dynamic Universe; the belief that it was static had been held for millennia and had not been questioned even by the great scientists and astronomers so far mentioned in this book. It led Einstein to remark that the cosmological constant, a repulsive force that he had introduced into his model of a static Universe in order to prevent it from collapsing, was no longer required. The relation between the velocity of galaxies and their distance is now known as Hubble's law and it shows that the velocity is proportional to distance, a property of expansion. The rate of expansion links the velocity to the distance and is known as Hubble's constant, one of the most important parameters in cosmology. The early value derived by Hubble was 150 kilometres per second for every million light years; in other words, a galaxy one million light years farther away than another one will be travelling at a speed 150 kilometres per second faster.

One of the many greatly significant results of Hubble's work was that his constant provided the very first indication of the age and size of the whole Universe. Since the Universe is expanding, it is possible to extrapolate back in time to the point where it would have negligible dimension – its starting point, now known as the Big Bang. From the expansion rate determined by Hubble, the age of the Universe came out as 1000 million years, but, as will be revealed later in the book, his distances were underestimated (like all early distance measurements in astronomy) and the current value for the age of the Universe is nearer to 10000 million years. But this estimate and the value of Hubble's constant is still uncertain today to about a factor of two, and is a matter of extensive study.

Of course, the extrapolation backwards in time, using the expansion rate measured today, cannot give an accurate estimate of the age (and therefore the size) of the Universe, because the expansion is against gravity, which must be slowing it down with time; hence, the expansion rate must have been greater in the past and will become slower in the future. A simple but very good analogy is to throw a ball into the air; its initial velocity is high, but gravity will slow it down to the point where it will stop and return to the ground. However, if we could throw it fast enough, it would be able to escape the Earth's gravity completely and never return; for the Earth, this velocity is about ten kilometres per second. This same problem exists in present-day cosmology: is the expansion velocity of the Universe enough for it to continue to expand forever, or is it insufficient, causing it to re-collapse into a Big Crunch? This is one of the central questions in modern-day cosmology, but is has not, as yet, been answered with any certainty.

At this stage in the story, it is appropriate once again to consider Olbers' paradox. Olbers had argued on the basis of a static Universe in concluding that it must be finite in space or in time. An expanding Universe could be infinite in space but it would be finite in time, namely the time taken since its expansion started. But an interesting philosophical point is that, in an infinite Universe (and we have no evidence to discard this possibility), the Earth would appear to lie at the centre, because we would only see that part of an infinite Universe to the distance that light can travel during the age of the Universe. This is not, as it seems, an anti-Copernican view, because any other observer on any other planet in any other galaxy, no matter how distant, would also seem to be at the centre of its Universe, but it would be another part of the whole infinite Universe. The discerning reader may now ask, "But what in the far future when the Universe becomes very much older?". The answer is that any observer in any part of the Universe will still appear to be at its centre, but the sky will still remain dark as long as the Universe continues at anything like its present expansion rate, because the high redshift of its most distant parts will make it appear to be cold. Olbers' paradox – "Why is the sky dark at night?" – had been fully answered, but in a way he could never have imagined.

The astronomical advances in the first part of the twentieth century were immense. As the clouds gathered before the outbreak of the Second World War, the energy source of the Sun and stars had been identified, the size and structure of the Milky Way galaxy had been determined, and the nature and limits of the whole Universe had been determined realistcially for the first time. The stage had been set for tackling the basic problems of the evolution of stars, of galaxies and of the Universe itself. But these great human intellectual achievements had now to pause, and not until the completion of the most extensive and destructive war in history, during 1939–45, did they resume, but with some of the most exciting and explosive developments that took astronomy into a new and golden age.

PART III

MODERN ASTRONOMY

In the period following the Second World War, technologies that had been developed for military purposes during that war were applied to observational astronomy with an impact that was possibly even greater than that caused by the invention of the telescope. For the first time ever, observations could be made beyond the very narrow confine of the optical spectrum, and the Universe could also be viewed in the radio, infrared, ultraviolet, X-ray and gamma-ray regions of the electromagnetic spectrum. The consequences of this were so exciting as to take astronomy into a new golden age which is still with us.

CHAPTER NINE
The New Astronomies

For I dipt into the future,
far as human eye could see,
Saw the vision of the world,
and all the wonder that would be.
Tennyson

Since astronomy is an observational subject, the techniques by which the Universe is observed are of crucial importance. However, the instruments employed today are highly technical and the methods of detection extremely sophisticated, putting a detailed presentation of them beyond the scope of this book. This chapter will therefore be confined to a broad overview of the modern techniques that are used today in the observation and study of the Universe.

After Galileo's seminal observations of the Moon, planets and stars, the telescope has been constantly increased in size and improved in quality, even to this day. In the nineteenth century, it was joined by two other developments of epoch-making importance, as has already been related. The first was the invention of the photographic plate which, with its ability to integrate the light over time, was able to out-perform the human eye and, additionally, left a permanent record of the observation. The second was the introduction of the spectrometer, which spreads light out into its different wavelengths and whose analysis has become one of the greatest tools in astronomy. In the immediate years after the Second World War, there were several large optical telescopes on various sites throughout the world, the largest of which was the 200-inch (5 metre) telescope on Mount Palomar in California, which was started before the Second World War but not commissioned until after it ended. Since then, the stimulus provided by the development of the new astronomies, the subject of this chapter, led to several countries building large modern telescopes, which are located in both the Northern and Southern Hemispheres, thereby giving a full coverage of the sky. They are placed on carefully selected sites in order to exploit the best possible astronomical observing conditions which, in general, are high mountains remote from densely populated regions; mountains are chosen in order to reduce the obscuring and blurring effects of the Earth's atmosphere and, in many cases, they lie above the normal levels of cloud cover; remote areas are chosen so as to avoid the visual pollution of the sky by artificial street lighting, which can, and does, limit the detection of the faintest objects in the sky. The largest of these new optical telescopes is located at a height of 14000 feet (more than 4000 metres) on the top of the extinct volcano of Mauna Kea in Hawaii (with several other optical, infrared and radio telescopes) and has a primary mirror of 10 metres in diameter. It is called the Keck Telescope, named after its private benefactor, and follows a great American astronomical tradition that is almost unique to the USA.

At the time of the ending of the Second World War, the whole of astronomy, which had come a long way, had depended entirely on observations in a very narrow window of the Earth's atmosphere, which was only an octave wide and which ranged from the near ultraviolet to the near infrared. This range of spectral transparency in the Earth's atmosphere embraced the range of response of the human eye (not by chance but as a result of natural evolution) and, since it lies in a region where the Sun's output is near maximum, it affords the route by which the Earth is illuminated and heated by the Sun. There is only one other very clear window in the Earth's atmosphere and this lies at radio frequencies; a few cracks occur in the infrared, but all other wavelengths of electromagnetic radiation are heavily absorbed and do not penetrate the Earth's atmosphere.

Two wartime technological developments, made entirely for military purposes, presented a possible basis for extending the range of astronomical observations from its previous confinement to the visible spectrum to other spectral regions. These two new technologies were radar, developed as a defensive tool to detect incoming enemy aircraft, and the propulsive rocket, developed as an offensive tool to deliver explosive warheads. The former presented the techniques for exploring the radio-clear window of the Earth's atmosphere, whereas the latter offered the possibility of launching equipment above the atmosphere, thereby avoiding its absorption and opening up all the other spectral regions to observation. Many visionary scientists, several of whom were physicists or engineers with no previous experience in astronomy, recognized the great potential of these new technologies for astronomy, and their pioneering efforts led ultimately to the Universe being opened to observations over the whole range of the electromagnetic spectrum. In addition to traditional optical astronomy, we now have radio astronomy, infrared astronomy, ultraviolet astronomy, X-ray astronomy and gamma-ray astronomy, to which we can also add neutrino astronomy. These new astronomies were to have an explosive and revolutionary impact on the subject.

174

☆ ☆ ☆

The first of the new astronomies was radio astronomy. It had been known that radio emissions came from space since 1931, when Karl Jansky (of Bell Telephone Laboratories in the USA), detected "noise" in a radio antenna he was developing for communication purposes. He noticed that this appeared every sidereal day (i.e. the time it takes the Earth to rotate against the background reference of the fixed stars). This is slightly shorter, by about four minutes, than a solar day, because the revolution of the Earth about the Sun adds an additional rotation as seen by the Sun. The recurrence of the noise on a sidereal day therefore meant that the source was a cosmic one and that it was located somewhere on the celestial sphere. (We now know, as Jansky did not, that the radio emission was coming from the centre of our Milky Way galaxy.) But it was the wartime development of radar that allowed the full opening of radio astronomy, because it extended the range in the frequency of radio waves that could be detected by a factor of more than a hundred, and in so doing, it turned out, took it into the important range for astronomical observations.

The aim of radar development was to generate a radio beam that could scan the horizon and, after bouncing off any approaching hostile aircraft, its reflection could be detected and the position and direction of the aircraft determined. The radio waves could penetrate cloud and the system would therefore give an early warning in all weathers and allow the appropriate defensive action to be initiated. However, there was a major technical problem: the highest frequencies of radio waves that could be generated and detected gave wavelengths as long as 10 metres, which was greater than the dimensions of aircraft and comparable only to their wingspan. Consequently, the reflected signal would be very small and difficult to detect and, further, the size and shape of the reflecting object could not be determined and its identification would be highly uncertain. In a remarkably successful radar development programme, the frequency of radio waves that could be generated and detected had been increased by more than a factor of ten in 1940, giving wavelengths of about one metre and allowing the establishment of an operating radar system, which proved to be one of the decisive factors in the Battle of Britain. By 1942, wavelengths down to about 10 centimetres had been achieved. It is not surprising that some of the earliest major initiatives in the post-war development of radio astronomy came from scientists who had worked on the British radar programme, notably Bernard Lovell at Jodrell Bank and Martin Ryle at Cambridge.

The instrumental challenges in radio astronomy stemmed from the same root as those in radar. Higher-frequency detectors were always needed in order to extend the range of astronomical observations to shorter wavelengths. Such detectors, or receivers as they are more commonly called for radio waves, respond directly to the incoming signal and are therefore able to measure phase as well as amplitude of the electromagnetic oscillation, a very important property, as will be seen shortly. The ability of radio receivers to respond directly to a radio signal means that they are coherent detectors which can be tuned to any one frequency, thereby avoiding any noise and confusion caused by other frequencies; these can themselves be observed by scanning the receiver through them. Research and development to increase the frequency capability of radio receivers has continued until today when millimetre wavelengths and even submillimetre wavelengths can be detected.

The greatest problem that faced the radio astronomers was that radio wavelengths are very long, particularly when compared to visible light, on which all previous astronomical observations had been based. This is because the wave nature of electromagnetic radiation places a fundamental limit on the image quality that can be achieved, even by a perfect telescope; this is given in angular terms by the ratio of the wavelength to the diameter of the telescope. For the early radio telescopes, this limit was about one, more than a thousand times poorer than that available with optical telescopes. This made the early identification of radio sources with their optical counterparts very difficult, because there were always many optical candidates in the field that embraced the possible location of a radio source. This problem was tackled with great ingenuity; the solution, conceived by Ryle, was to build several radio dishes which were separated by considerable distance (such as a few kilometres) and link the signals from each of those into an interferometer which combined them. Although the total collecting area of such an array was only the sum of the areas of the individual dishes, the image quality was determined by their separation, with a consequent major improvement in image quality. The largest such single system is in New Mexico and is called the Very Large Array (VLA); it is many tens of kilometres in extent and consists of 27 dishes of 25 metres diameter. (The earlier Lovell telescope at Jodrell Bank which aroused such great public interest, is a single dish of about 75 metres diameter.) This interferometric combination of radio signals was done in a way quite different from optical interferometers, where the two beams were physically brought together in order to interact. In the radio case, the amplitude and phase of the signals from the individual dishes were recorded and their combination and interference was carried out numerically in a computer. This allowed the principle to be extended to even wider separations and allowed the linking of radio telescopes located in different continents. Such intercontinental baselines gave a remarkable accuracy of a few milliarcseconds – a penny coin as seen from a thousand kilometres! Hence, radio astronomy had responded to its greatest challenge of low angular resolution by reaching an accuracy far in excess of that which had been achieved in optical astronomy.

The first astronomical observations to be conducted from above the Earth's atmosphere were made with equipment carried in sounding rockets, which were launched vertically upwards and could spend a few minutes free of atmospheric absorption before returning to the ground. The first celestial body to be observed in hitherto forbidden regions was the Sun, whose spectrum, over several launches, was recorded from the atmospheric cut-off in the near ultraviolet down to the X-ray region. The use of sounding rockets was then extended to the study of other cosmic objects such as the bright stars, thereby laying the foundations for ultraviolet and X-ray astronomy. These pioneering experiments in space astronomy started in the USA (the very first used captured German V2 rockets) and were later extended by groups in other countries, but it should be acknowledged that the opening up of this new area owed much to the ability and vision of two scientists, Herbert Friedman and Richard Tousey, both of the US Naval Research Laboratory near Washington.

The sounding rocket has many advantages for making astronomical observations from above the Earth's atmosphere. It was relatively cheap, payloads could be prepared quickly, and data could be recovered on return to the ground (e.g. on photographic film), without the need for telemetry. But it had one overriding disadvantage: observing times were limited to several minutes at most, whereas astronomy needed long observing times in order to detect faint objects. The only answer was that space observatories had to be launched into Earth orbit, where they could have lives of a year or more before return to Earth, because of atmospheric drag. This posed a range of new challenges which were common to all the future space astronomies – ultraviolet, X-ray, gamma ray and infrared. Each of these had quite different instrumental problems, which will be discussed shortly, but all had to meet the same general problems which are presented by the need for a space observatory to be built and tested on the ground, placed into the nose cone of a rocket and then launched into space, where it had to conduct its observations over a considerable period of time. These common problems can be broken down into three general areas. The first is the constraints imposed by the launch vehicle, which place severe limits on the volume, weight and on-board power available to the payload. The volume and shape are strictly constrained by the shroud of the rocket head, into which the equipment has to fit; this results in design layouts which are different from those appropriate for a ground-based observatory, but have to be optimized within the limits set by the launcher. The payload weight is also strictly limited by the power of the rocket, which can only launch a certain weight into any required orbit; this necessitates the design of lightweight structures, but still with the stability requirements required by the system. The on-board power limitation is a direct result of the limit on payload weight, because the satellite has to generate its own power and it does this by converting solar radiation into electricity by means of photoelectric cells assembled on solar paddles; these have to be built within the weight budget and therefore they limit the power available; the consequence is that the satellite has to be designed so as to operate at the lowest possible power levels.

The second area of problems to be faced by any orbital space observatory is posed by its many different environments. It has to be designed within all the constraints of weight and shape, built, assembled, tested and commissioned on the ground, at atmospheric pressure and in the Earth's gravitational field; the tests have to cover the conditions it will meet in its subsequent environments. It will be placed in the nose cone of a rocket and subjected to a high acceleration and, worse, severe vibrations, which it must survive, despite its low-weight structure and sophisticated detectors. Finally, it is placed in a force-free orbit in a vacuum environment where its high-voltage components, usually required for the detectors, must not break down.

The third area of problems to be met by a space observatory is posed by its remoteness; repairs and replacements are not possible, so the whole system has to have a degree of reliability beyond anything ever considered for ground-based studies. This led to the development of "space qualified" components, which had been subjected to extensive tests under vibration, in vacuum, at different temperatures and for long operational lifetimes. An additional approach is to introduce redundancy into the design, that is to build two components into the system when only one is needed, together with a switch that will bring in the redundant component if the prime component fails; of course, the reliability of the switch also has to be a factor in assessing the most reliable design. A more subtle aspect of orbital operations is the thermal environment. The spacecraft is heated by the Sun on one side only and this will cause temperature gradients which may be undesirable for some functions such as precisely aligned optical components; the spacecraft also generates heat from the power it uses to conduct its operations. These factors make it necessary that a thermal design is conducted as well as optical, mechanical and electronic designs, in order to ensure that the spacecraft will be within the temperature range and temperature gradient required for the optimum performance of its components. Methods of thermal control are rather limited, but passive ones include a choice of external paint, which can reduce or increase the fraction of the incident solar flux that is absorbed as heat, and a judicious distribution of the internal heat sources where possible.

A final problem caused by operating in space is that of communication; instructions have to be transmitted to the spacecraft, and the data it collects have to be returned to the ground. Radio waves of appropriate frequency are selected and ground stations built which can both transmit to and receive from the satellite. For near-earth orbits, which cover the majority of cases, one ground station has only one period of contact of several minutes during Earth orbit, which lasts typically about 90 minutes. During that period of contact, sufficient instructions have to be sent to the spacecraft to enable it to operate for at least another orbit, and all of the data it has accumulated have to be transmitted and received by the station. The spacecraft therefore has to have the capability to function automatically, often carrying out complex operations, and to store its data on board for at least the period between two successive radio ground contacts. These requirements, which add significantly to the complexity of the satellite, can be greatly eased by adopting a geosynchronous orbit, whose period is the same as the rotational period of the Earth (a sidereal day) when the spacecraft is in continuous contact with a single ground station; such an arrangement has been exploited by the International Ultraviolet Explorer (IUE), an observatory satellite launched in 1978.

The many problems caused by operating from an orbiting satellite platform, listed above, are common to all space activities and not just those of an astronomical nature. But one particular challenge which, if not restricted to astronomy, is far more demanding in that area than any other, is the need to point to any celestial object in the sky and to lock on to that object for considerable periods of time with a very high precision of about one second of arc or even better. This requires attitude control systems of great sophistication which can slew the telescope to any point in the sky, locate the object to be observed and then track it for the period of time needed to study it. The standard solution is to use gyroscopes to give the spacecraft a basic inertial reference, but ultimately an astronomical reference has to be used, given by measuring the position of the object itself or a nearby star. The fact that space observatories have been launched with the ability to locate and track the faintest of astronomical sources represents a fine technological achievement.

In addition to the general problems faced by operations in space, the several spectral regions opened up to astronomical observations presented instrumental challenges as different and varied as those between the two ground-based activities of radio and optical astronomy. In ultraviolet astronomy, techniques were somewhat similar to those in optical astronomy, with telescopes and spectrometers of a familiar design.

However, special coatings had to be developed for the mirrors, since most materials do not reflect strongly in the ultraviolet; it was also necessary to find quality optical materials for high-transmission optics, like the window of a detector, since very few are transparent to ultraviolet radiation. These developments allowed efficient operation for most, but not all, of the ultraviolet region of the electromagnetic spectrum; in the very far ultraviolet, windowless detectors have to be used and grazing incidence optical techniques developed, in which reflections are constrained to glancing angles from surfaces.

Detectors for measuring ultraviolet radiation were also similar in principle to those used at optical wavelengths but, of course, photographic films, which had been used in many sounding rocket experiments, could not be used in satellites and were replaced by electronic detectors whose basic function was to detect an incoming photon by making it release an electron from a surface and then accelerate it to a high enough voltage to be detected. The image being recorded can then be assembled by counting the number of photoelectrons in each picture element (pixel). These photon-counting imaging sensors, together with solid state detectors called charge-coupled devices, have been used in space and on the ground, covering a spectral range from the far ultraviolet to the near infrared. Their performance is such that, except for specialist purposes, they have also replaced the photographic plate in ground-based optical astronomy.

In X-ray astronomy, the instrumental challenges are very different from those posed by ultraviolet and optical astronomy. X-ray photons have a much higher energy, equivalent to electrons that have been accelerated to about a thousand volts compared to about 5 volts for ultraviolet photons, and are therefore easy to detect. However, X-rays penetrate materials rather than be reflected by them, and therefore cannot be collected and imaged by telescopes in the conventional sense. The earliest payloads therefore used a mechanical method of observation consisting of several thin cylinders, mounted in parallel, which allowed X-rays from only a small region of the sky to be viewed. The resolution of about one degree was very crude by optical standards, and the early stages of X-ray astronomy were marked by one of the same problems as that faced in the early stages of radio astronomy, namely to identify any source with its optical counterpart. The detectors were simple and effective consisting of gas cells sealed by thin windows, through which the X-rays could penetrate and, on passing through the gas, ionized the atoms in its path; the released electrons were then accelerated to the walls by a high voltage and their signal was detected and measured. Since the more energetic photons would liberate more electrons, the height of the signal pulse was a measure of photon energy (and therefore wavelength). The spectral resolution that this afforded was not great (about the same as the eye in separating the colours of the rainbow) but was nevertheless very valuable. As the importance of X-ray astronomy grew, technological developments led to the building of X-ray telescopes with accuracies of a few seconds of arc. These were grazing-incidence telescopes, based on the principle that even X-rays would be reflected by a surface if they struck it at a sufficiently low glancing angle, like the distant reflections seen from a road surface while driving a car. This is a very sophisticated kind of telescope, presenting major challenges in terms of surface quality, but such systems have been developed and launched successfully.

In gamma-ray astronomy, no kind of imaging telescope is possible because no material can reflect such radiation even at the finest of glancing angles. However, the photon energies are so high, from hundreds of millions to thousands of millions of electron volts, that they ionize the gas in a large chamber along the whole length of their path, allowing the energy and direction of the photon to be determined.

The overview of the new astronomies outlined above was given in the chronological order of their development – radio, ultraviolet, X-ray and gamma-ray astronomy. The final one, infrared astronomy, was last because it had to face the most severe instrumental problems of all for two basic reasons. First, infrared detectors are not as sensitive as detectors in the neighbouring radio and optical regions of the spectrum; the former are coherent detectors, which can tune directly to the frequency of the incoming radiation, whereas the latter are quantum detectors which can detect individual photons. In the infrared, the frequencies are too high for the radio technique and the photons have insufficient energy for the optical technique; infrared detectors therefore have to measure the amount of energy deposited by the incident signal against the background of all other inputs to its surface. The second, and more severe problem, is the fact that, at normal operating temperatures (in the neighbourhood of 300° Kelvin, which is 27° centigrade and 81° Fahrenheit), all materials are emitting strongly at infrared wavelengths, causing the equipment itself to swamp the faint astronomical signals. An analogy with optical astronomy, which embraces the visible region of the spectrum to which the human eye responds, is as if the telescope, instruments and detectors were white hot! The solution was very easy to identify but extremely difficult to implement: the equipment had to be operated at sufficiently low temperatures that its self-emission became negligible in the infrared. This required it to be encased in a surrounding chamber containing liquid helium, whose evaporation temperature is particularly low at 3° Kelvin, that is 3° above absolute zero where the atoms no longer have any kinetic energy of motion and therefore no ability to emit radiation. This kind of extremely low-temperature (cryogenic) operation presented additional challenges such as the thermal shielding of the liquid helium jacket, the operation of the equipment at such low temperatures, and the minimization of its power consumption to extremely low levels. With these precautions to minimize the heat input to the liquid helium from both external and internal sources, the first Infrared Astronomical Satellite (IRAS, a Dutch-led mission involving the USA and the UK, launched in 1983) was able to operate for ten months before its liquid helium boiled away.

The great success of the new astronomies – radio, infrared, ultraviolet, X-ray and gamma ray – was attributable to their attraction as new and exciting fields to scientists and engineers of the highest quality, who had the vision to recognize the great potential of extending the range of wavelengths accessible to observations of the Universe. As a result of their endeavours, the post-war years mark the most dramatic period in observational astronomy since 1609, when Galileo first pointed his telescope at the sky. The expansion of the observable spectrum from about one to fifty octaves led to a succession of major discoveries that have revolutionized human knowledge of the nature and origin of the Universe. These will be revealed in Chapter 12, and their impact on the various areas of astronomy, such as the Solar System, stellar evolution and cosmology, will be described in the appropriate chapters.

CHAPTER TEN

Probing the Solar System

And pluck till time and times are done
The silver apples of the Moon
The golden apples of the Sun
Yeats

The development of the rocket allowed the launch of space platforms from which astronomical observations could be made in those regions of the spectrum that were obscured by the Earth's atmosphere. But, in addition to their use as observing platforms, space vehicles could also be used as probes to take equipment to extraterrestrial bodies in order to make measurements *in situ*. The first of these was the remarkable American Apollo mission, which took men to the Moon and then returned them with a selection of lunar rocks for laboratory analysis. This was a technological feat of the highest order and it had a major impact internationally, but its astronomical returns can only be described as interesting. Quite contrary to this were the results of interplanetary probes by the USA and the (then) Soviet Union, notably the immensely successful *Voyager* probes, which were to revolutionize knowledge of the local space environment – the Solar System.

In the post-war years, a series of planetary probes visited and made close encounters with every planet except Pluto, and in two cases, Venus and Mars, vehicles were landed on the surface to make measurements. The results received great public acclaim and were greatly disseminated in the newspapers, television and popular and scientific articles. There is therefore no need to repeat these findings in any great detail here; instead, a broad survey will be given, so as to give the reader a feeling for the scale and nature of the bodies in the Solar System, including the Sun. However, one particular question will be addressed in some depth, a question of immense human interest and significance, a question which, like the nature of the Universe, has been asked since ancient times: Is there life anywhere else in the Universe other than on the Earth? Or, in more personal terms, are we alone? The starting point in considering this question is obviously the planets.

Five planets (six with the Earth) have been known since ancient times; in order of distance from the Sun, they are Mercury, Venus, (Earth), Mars, Jupiter and Saturn. A sixth was added in 1781 when an English musician and amateur astronomer, William Herschel, discovered a new planet with the aid of a superb telescope which he had built himself. He named it Uranus after the mythological father of Saturn, beyond which it revolved. Observations of Uranus between 1790 and 1840 revealed that its orbit was not a perfect ellipse but was subject to perturbations, which were proposed to be caused by the gravitational interaction with an unknown, more distant planet. In 1845, John C. Adams in England and Urbain Leverrier in France independently calculated the orbit of the new planet and were able to predict its position. Although the observatories in their own countries were among the best in the world and were the first to receive these predictions, it was not until Leverrier passed them on to Berlin that Johann Galle discovered the new planet within one degree of the predicted position. It was named Neptune after the god of the sea. The last and outermost planet was discovered in 1930 by Clyde Tombaugh at the Lovell Observatory in Arizona. Its discovery again followed predictions based on perturbations in the orbit of Uranus and it was named Pluto after the god of Hades.

A pictorial overview of the Solar System is given in Colour Plates 8 and 9. The first shows the sizes of the planets and the Sun, again to scale, together with some aspects of their characteristics, including their colour. The second gives the orbits of the planets about the Sun to scale. For the reader who also likes numbers, these are listed in the accompanying table and are given in Earth units for distance, size, mass, and so on. The range in distance is very considerable, the outermost planet, Pluto, being a hundred times farther from the Sun than the innermost planet, Mercury. The total extent of the Solar System, defined by the diameter of the orbit of the most distant planet, Pluto, is 12000 million kilometres or, in light travel times, 11 light hours. Because of Kepler's third law, the periods of the planets vary even more, by a factor of a thousand, than do their distances from the Sun. The range in mass is even greater, with the heaviest, Jupiter, being 100000 times more massive than Pluto, 5000 times more massive than Mercury and more than 300 times more massive than the Earth.

The planets can be placed into two groups: the terrestrial planets, comprising Mercury, Venus, Earth and Mars; and the giant planets, comprising Jupiter, Saturn, Uranus and Neptune. Less is known about Pluto than any other planet, but what is known shows it to be anomalous in so many ways that it should be considered separately from the other bodies of the Solar

The main parameters of the planets expressed in values relative to the Earth.

Name	Distance from Sun (AU)	Period of revolution	Size (compared to Earth)	Mass	Mean density
Mercury	0.39	88 days	0.38	0.06	0.99
Venus	0.72	225 days	0.95	0.82	0.95
Earth	1	1 year	1	1	1
Mars	1.5	1.9 years	0.53	0.11	0.71
Jupiter	5.2	12 years	11	318	0.24
Saturn	9.5	29 years	9	95	0.13
Uranus	19	84 years	4	15	0.24
Neptune	30	165 years	4	17	0.30
Pluto	40	249 years	0.2	0.003	0.37

Earth distance = 1 AU (astronomical unit) = 150 million kilometres = 8.3 light minutes
Earth diameter = 12800 kilometres
Earth mean density = 5.5 × water

System. Its orbit does not lie in the narrow band of the ecliptic plane that contains the orbits of the other planets, but is inclined to it by 17°; its orbit is also the most elliptical of all the planets and, at closest approach, it falls within the orbit of Neptune where it is now and will remain until the year 2000; it is also the lightest and smallest of the planets. These factors raise the question as to whether Pluto had a different origin from the other planets and whether, for example, it was originally a moon of Neptune which was detached by some gravitational encounter; but, at the present time, the matter is an open question.

The terrestrial planets are the four closest to the Sun, are relatively small, have high mean densities and are rocky in nature, a result of the chemically selective process that caused their formation. The giant planets are farther from the Sun, have low mean densities and are gaseous in nature, a result of their being formed from the solar proto-nebula, which was composed mainly of hydrogen and helium. The chemical composition of the terrestrial and the giant planets is therefore very different; the latter were large enough to capture gravitationally most of the elements in the collapsing medium, whereas the former were too small and their condensation needed the chemical processes that allowed the heavier elements to combine into molecules, then into compounds and then to coalesce, while most of the abundant hydrogen and helium escaped.

Most of the planets are not single bodies but have their own satellites, or moons, trapped in their gravitational fields and orbiting them in the same way as the planets orbit the Sun. The Earth has one moon, Mars has two and even Pluto has one; only Mercury and Venus have none. The giant planets have several moons and are also surrounded by complex ring systems, of which Saturn's are the most striking. The four moons of Jupiter, observed by Galileo in 1610 and later named as Io, Europa, Ganymede and

Callisto, have been increased to 16, first by improved telescopic observations and, more latterly, the two superb American *Voyager* missions during their flypast of the planet. During close approach, the *Voyagers* made the most detailed observations ever of Jupiter and its moons; the results lie outside the main theme of this book, but one example will be given. Io, the closest moon, was shown to be volcanically the most active body in the Solar System and this is believed to be the result of its tidal heating by Jupiter. In addition to lava flows, gases composed of elements such as hydrogen, oxygen and sulphur escape from Io's tenuous atmosphere and enter its orbit, where they are ionized by the Sun's ultraviolet radiation field. On ionization, they are immediately swept up by Jupiter's magnetic field, which is nearly 20 000 times more intense than the Earth's and which is moving at the high speed of about 7 kilometres per second because of the rapid rotation of the planet at once every ten hours. This acceleration of Io's volcanic output creates a high-temperature plasma in the region of its orbit, called the Io torus; this decays with time into Jupiter's magnetic field configuration and causes radio emissions that have been detected for some time on Earth.

There are other bodies in the Solar System other than the Sun and planets. These include the asteroids, of which several thousand are now known and catalogued; they orbit the Sun between Mars and Jupiter, with periods varying between about three and five years. They are rocky bodies with typical diameters between 10 and 100 kilometres, although the largest, Ceres, is about a thousand kilometres; they may have been a planet that was broken up or prevented from forming, by the tidal forces of Jupiter. The asteroids, like the planets, lie in or near the ecliptic plane, but the comets, on the other hand, come in from all directions and follow highly elliptical or parabolic orbits. They come from the Oort cloud, named after the Dutch astronomer who proposed its existence, and which surrounds the Solar System at about one light year's distance; it contains many potential comets which, since the nearest stars are only a few light years away, can be perturbed sufficiently to fall into the gravitational field of the Sun. If they are subject only to the Sun's gravitational field, they will follow a nearly parabolic orbit and return to the Oort cloud not to be seen again; however, if they suffer gravitational perturbations by the planets on their initial infall, they can become trapped in the Solar System with highly elliptical orbits. The most famous of these recurrent comets is Halley's, which has a period of 76 years, and was first observed in 240 BC. One of its returns coincided with the momentous Battle of Hastings in 1066, thereby contributing to the illusory belief that comets are heralds of misfortune.

In addition to the comets and asteroids, there are many smaller bodies called meteors and meteorites present in the Solar System, whose origin is probably the break-up of larger bodies such as the asteroids and comets, caused by gravitational perturbations by the larger planets. They are the main contributors to the 200 000 tonnes of cosmic material that impacts and enters the Earth's atmosphere every year and they are easily visible as "shooting stars" when they burn themselves up because of their high velocities. They are quite harmless, but an important and interesting question, which is worth a digression at this point, is the extent to which the impact of external bodies on the Earth becomes a hazard to life. The Earth's atmosphere presents a very formidable shield; even objects up to about a tonne in weight are broken up into small particles, which produce meteor trails as they burn up. Although a one tonne meteorite or asteroid will cause a great visual spectacle of shooting stars, it will be of no danger to life on Earth. The question is, what level of mass and energy has to be reached before a celestial object will penetrate the Earth's atmosphere and cause major local damage, and at what level will that damage become global in nature. At this point, we have to consider the energies of those impacting bodies, which are enormous because of the very high velocities involved, typically about 20 kilometres per second for asteroids and ranging up to 60 kilometres per second for comets. The energy of the body scales linearly as its mass but as the square of its velocity; hence, the fastest comets have about ten times the energy of a typical asteroid for the same mass. The energies involved are so high as to require some unit of measure that is meaningful to the reader. This has been provided by the development of nuclear weaponry, where the power of bombs is expressed in their equivalence to the number of tonnes of conventional high-explosive TNT. The greatest bomber raids of the Second World War dropped about 1000 tonnes (1 kilotonne TNT), and the two atom bombs dropped on Hiroshima and Nagasaki were of 15 and 20 kilotonnes TNT respectively. The unit to be adopted here is a million tonnes (megatonne) of TNT; to give a feeling for how large this is, the total energy of all bombs dropped by all countries during the six years of the Second World War was three megatonnes.

The Earth's protective atmosphere burns up nearly all of the cosmic objects that strike it, producing only spectacular showers of shooting stars. Even bodies of a few tens of metres in diameter and energies of a few megatonnes of TNT would not penetrate to the ground. But clearly there must be some threshold in size and energy above which a cosmic body would penetrate the atmosphere, reach the Earth's surface and cause severe local devastation. This threshold cannot be estimated with any precision but is probably reached for objects of about a hundred metres in diameter and about ten megatonnes of TNT in energy. Such occurrences are very rare but do happen occasionally, the last being in 1908 when a massive meteorite struck a forest in Tunguska, Siberia, with an energy estimated at 20 megatonnes of TNT. Trees were completely felled over an area of a thousand square kilometres, and fires were ignited by the central fireball. Another known impact occurred about 50 000 years ago and left a crater of more than a kilometre in diameter in the Arizona desert. There will have been several other such events which have not been recorded because there is no geological record, such as those hitting the oceans. But it is clear that such impacts are very rare and, although statistics can be very misleading with such small numbers, you will get some feel for the probability if I suggest an average occurrence of about one in every thousand years. Since any such impact will cause severe damage only locally, it will endanger human life only if it strikes an urban area, when the effects will dwarf other natural disasters. But the fact that urban areas cover a minute part of the Earth's surface, together with rare occurrence of such impacts, means that this should not be regarded as a problem of major concern.

The next question is – at what level of size and energy would a cosmic projectile cause not only disastrous local effects but also major global effects on life by disturbing the whole world climate? The answer depends on the nature of the object, its angle of entry and its point of impact, but even if these were known, only very rough estimates could be made. Clearly, for projectiles of energy greater than the 1908 Siberian meteorite, the damage will be correspondingly greater, but studies have concluded that a point would be reached when the release of dust into the atmosphere, caused by the impact, would reach such a level as to significantly shield the Earth's surface from solar heating, with a consequent reduction in global temperatures of several degrees centigrade over many months. This would lead to massive crop failures throughout the world, and have other major effects on life as to make it a disaster of historic proportions. But it does require a cosmic projectile of enormous energy, more than about one kilometre in size and approaching 100 000 megatonnes of TNT in energy. To put this in context, the total energy of nuclear arsenals of all the world's nuclear powers is estimated to be one-tenth of this value. Nevertheless, studies of a consequence of the unthinkable, a worldwide nuclear war, have suggested the creation of a "nuclear winter".

All this leads to the critical question of the likelihood of such an Earth-shattering cosmic event happening, and the short answer is that it is very, very remote. In the early stages of the Solar System, the frequency of cosmic impacts was much greater than now, as can be seen by the many impact craters on Mercury and the Moon, two bodies without an atmosphere, whose craters are therefore preserved and not eroded by weather as they are on Earth. Most of the early meteorites and other bodies have now been swept up by the planets, notably Jupiter, with its large cross-section and high gravitational field. But public attention on this issue was revived by the collision in the summer of 1994 of the Comet Shoemaker–Levy with Jupiter (comets are always named after their discoverers, in this case two astronomers). The energy was an astonishing 100 million mega-tonnes of TNT, or more than a thousand times the estimated entire man-made nuclear arsenal. The cometary fragments (it broke up previously because of Jupiter's gravitational field) entered the gaseous planet and dissipated their energy by thermal heating and tidal effects on a global scale. Such an impact on Earth would cause a doomsday, with the probable elimination of civilization and the possible extinction of the human race. But the chance is extremely remote; the last time such an energetic impact occurred on Earth is believed to have been 65 million years ago, when the dinosaurs and some other life species were eliminated in its aftermath. This belief has been greatly strengthened by recent researches, which have identified the fossil remnant of a massive crater formed at about that time and centred near the shoreline of the Gulf of Mexico in the Yucatan Peninsula. It has since disappeared as a result of the combined effects of erosion and deposition, but its remnant was detected by the application of sensitive instruments to measure the local magnetic and gravitational fields; these allowed the crater to be mapped and they yielded a diameter of 180 kilometres (113 miles). Such an immense crater implies an impact energy approaching 100 million megatonnes of TNT, placing it in the global catastrophe domain and making it the most likely cause of the extinction of the dinosaurs, thereby explaining a longstanding puzzle. Today the main source of a possible major cosmic impact on Earth lies with the many asteroids, the most massive of which do have energies in the global catastrophe domain, but the chance of an Earth collision is so minute as to be negligible; the fact that triggers any discussion of such a possibility is only the, literally, Earth-shattering consequences of such an event.

This chapter on the Solar System deals mainly with the planets, but the most dominant body in the Solar System is the Sun, which contains all but 0.1 per cent of the total mass. It is the only star close enough to be studied in detail, and observations show that its surface has considerable fine structure, with features that vary with time, some rapidly and unpredictably, and others more slowly and more predictably. These effects are important and interesting, but they represent only a small fraction of the Sun's energy generation. It is most appropriate to discuss them here, but the Sun as a star, including its energy generation and evolution, will be treated in the next chapter with all other stars.

Some of the most prominent features on the Sun's surface are areas that are relatively dark compared to the rest of the disk. Called sunspots, they were known to and observed by the ancient Chinese astronomers, and were "rediscovered" in 1610 by several people, including Galileo. Sunspots have been observed systematically since 1750 to the present day and the data show that their numbers vary on an 11-year cycle, reaching a peak and then declining to a level where they are almost absent, beyond which they reappear again. They are located at roughly fixed latitudes above and below the solar equator; they first appear at mid-latitudes, and then approach the equator as they increase in number, being closest to it when they disappear at the end of the cycle. Although the 11-year cycle is quite regular, the peak number of sunspots (and other aspects of solar activity) varies considerably. The development of spectroscopic techniques had led to one in which magnetic fields can be measured and this shows that the sunspots are areas of strong magnetic fields which emerge from the solar interior. Their typical strength is a few thousand times greater than the Earth's magnetic field, which is able to align a compass needle towards the north and south magnetic poles. A magnetic field exerts a pressure and therefore causes a local decrease in temperature within the sunspot as it reaches pressure equilibrium with the surrounding atmosphere; it is that temperature drop which makes them appear to be relatively dark. The magnetic measurements reveal that the 11-year cycle is actually a 22-year cycle in which the magnetic fields of sunspots completely reverse their polarity, pointing one way in the first 11 years (say outwards from the solar surface) and pointing in the opposite direction (inwards) in the second 11 years.

Other solar phenomena also follow an 11-year cycle in their frequency and are also dominated by the effects of the Sun's variable magnetic activity. Prominences rise above the disk and form loops which are clearly confined by a magnetic field, but sometimes the field cannot hold them and they erupt to escape into the interplanetary domain. The most violent events are the solar flares, which are believed to be the result of a catastrophic breakdown in the local magnetic field; they occur very rapidly, hurl out highly energetic particles and produce X-rays. All of these effects of solar activity are a result of the Sun's generation of magnetic fields, both locally and globally, but how these are generated, and their cyclical nature, are not understood in any detail, but they are certainly the result of the dynamic motions that occur in the solar interior.

The chance equality of the angular diameters of the Moon and Sun, as seen from the Earth, causes a total solar eclipse on those occasions when they happen to be co-linear. Such occurrences blot out the solar surface (called the photosphere because it is the source of most of the emitted light) and reveal the presence of an atmosphere lying above the Sun proper and stretching out to beyond a solar radius. This medium became a matter for extensive study in the nineteenth century, as has already been related, and eclipse observations revealed a bright inner portion (called the chromosphere because it is coloured) and a very extensive halo (called the corona). Also, the application of spectroscopy revealed the presence of several emission lines, many of which could not be identified at the time. It was proposed that the very strong yellow spectral line emitted by the chromosphere was attributable to a hitherto undiscovered element, which was named helium after the Sun. This was shown to be correct some 25 years later when helium was isolated in the laboratory in 1895. It was also proposed at that time that the several spectral lines emitted by the corona were caused by another undiscovered element called coronium. In this case, the proposal was wrong, but it took more than 70 years to establish this and to find the right answer.

In 1942, when the greatest war of destruction was being waged through the rest of Europe and indeed, most of the world, Bengt Edlen, a spectroscopist in neutral Sweden, was recording the spectra of highly ionized elements, that is, elements stripped of several electrons in very high voltage discharges. The observed spectra lay in the soft X-ray region and Edlen systematically catalogued these for interest only and without any thought as to their possible future use. But he was very perceptive and, in constructing the energy levels of electrons, he noticed that the small separations in the lowest levels of some species corresponded to the spectral lines emitted by the corona. The separations were not a surprise and were well understood from quantum mechanics (Ch. 7); small differences in the energy of an electron in a prime orbit could be caused by the flip of its spin from a parallel to an anti-parallel position compared to the nucleus, and other subtle effects. Hence, although Edlen had not observed the coronal lines directly, he was able to infer their wavelengths to such an accuracy and to obtain such excellent agreement as to make his identifications conclusive. The coronal spectral lines were attributable to extremely ionized species of common elements such as calcium (Ca), iron (Fe) and nickel (Ni); one of the strongest coronal lines, emitted in the green, is caused by iron which has been stripped of 13, or one half, of its 26 electrons. The astronomical conclusion was swift and obvious: the corona had to be very hot – about a million degrees Kelvin.

Although the corona is exceedingly hot, it is also very tenuous (like a hot vacuum), so that its total energy output is trivial (less than 1 per cent) compared to the Sun proper. That is why its existence was revealed only by the chance occurrence of total solar eclipses which blocked out the main energy output of the Sun. But this situation is inverted if the Sun is observed in X-rays where the photosphere is dark and the corona is emitting at its maximum frequencies. Such observations became possible in the immediate post-war years using sounding rockets, that is, rockets that could carry astronomical instruments above the obscuration of the atmosphere, albeit for only a few minutes. The very first observations in space astronomy were made from such sounding rockets and were concentrated on the Sun. From the far ultraviolet to the soft X-rays, the Sun's emission is attributable entirely to the chromosphere and corona, which show a rich spectrum of emission lines, whose analysis enabled the temperature structure of those media to be determined. Just above the photosphere, the temperature reaches a minimum of about 4000° Kelvin, but then climbs very rapidly through the chromosphere until it levels off at about two million degrees in the corona. X-ray images reveal the hottest areas and show that the corona is far from being uniform but has regions of high emission and of low emission, the latter being called coronal holes.

The mechanism by which the corona is heated is still not fully understood today, mainly because of the complexity of the phenomena involved. The energy source is certainly the dynamic motions that exist in the solar interior, some of which are coordinated in nature, probably as a result of the Sun's rotation, causing dynamo effects which produce magnetic fields; others are random in nature and are present as turbulence, which is driven by the outward flow of energy generated in the solar interior, and which are evident in the granulation of the Sun's surface where the tips of the turbulent eddies can be seen and are about a thousand kilometres across. The dynamo effects are complicated by the fact that the Sun's rotation is differential, having a period of about 27 days at the equator but slowing to nearly 30 days at high latitudes; this probably accounts for the sunspots, but their periodicity is still not understood. The turbulent zone is probably the source of the energy that heats the chromosphere and corona since it can generate waves of various kinds that propagate upwards and deposit their energy in the higher layers. One form of propagation is by acoustic or sound waves (if the reader wants evidence that a turbulent medium generates noise, listen to a jet engine, but remember that the frequencies in the Sun are much lower); as such waves propagate into the higher, more rarefied media, their velocity amplitudes will increase until they reach the velocity of sound, when they will deposit their energy in shock heating. This is a probable contributor to the heating of the lower chromosphere, but the upper chromosphere is mainly heated by thermal conduction from the exceptionally hot corona. So, the important question is, what heats the corona? The answer is still not resolved in detail, but it is almost certainly explained by some form of magnetohydrodynamic wave, which can carry mechanical energy through a magnetized, ionized medium; for example, transverse waves, unlike the longitudinal sound waves, can be carried, in which the magnetic field acts like a taught string being plucked. But the precise form of such a heating mechanism is still not understood.

The very hot nature of the corona is such that it cannot be contained by the Sun's gravitational field and it therefore evaporates, causing a solar wind in which ionized particles, mainly protons and electrons, stream out into interplanetary space at velocities of several hundred kilometres per second. The solar wind increases during the periods of highest solar activity, that is every 11 years, and can be intensified considerably by the eruptions of solar flares, which inject much more energetic particles into the wind and can cause effects on Earth such as radio blackouts. On reaching the Earth, the solar wind is largely deflected by the terrestrial magnetic field, but many of the particles diffuse into it and become trapped to form the Earth's magnetosphere, called the Van Allen radiation belts, after their discoverer. These particles leak into the Earth's atmosphere at the cusps of the magnetic field near the (magnetic) poles and cause the aurora borealis or northern lights.

Although the details of the formation of the Solar System are still quite uncertain, the broad picture of the process is generally accepted, that is, of gravitational collapse from the pre-solar nebula, an interstellar cloud composed mainly of gas and a little dust. From the estimated ages of the galaxy and the Sun (see Ch. 11), it is known that the Sun was formed about 5000 million years after the period of rapid star formation which marked the start of the galaxy and which occurred in the very early stages of the Universe. At that time, the primordial material was composed entirely of hydrogen (three quarters by mass) and helium (one quarter), with minute traces of lithium, beryllium and boron, but none of the other 87 heavier elements, from carbon to uranium, which are naturally present on Earth. These elements were made and processed by thermonuclear transformations in the high-temperature cores of the earliest stars, mainly the massive ones whose lifetimes are short and end in cataclysmic explosions which hurl their nuclear waste into the interstellar medium. This "waste" was the seed of life on Earth; all life forms, including we human beings, are composed mainly of those elements that were manufactured by earlier generations of stars, now dead or in the last stages of their lives. As I said in Chapter 1, we are children of the stars.

When the Solar System was formed, the medium from which it condensed contained 2 per cent of the heavier elements from carbon and above, some of which had coalesced into small dust particles. Having some angular momentum, the pre-solar nebula would be rotating and this would speed up (like an ice skater whose arms are contracted) and the nebula would flatten into a disk. The central blob collapsed to form the Sun, which had sufficient mass to heat its interior to a temperature high enough to ignite, cause thermonuclear burning and hence become a star.

Other contractions did not have sufficient mass to become stars and they formed the planets. The giant planets (Jupiter, Saturn, Uranus and Neptune) condensed from perturbations large enough to confine all the nebular material and are therefore formed, like the Sun, of about 73 per cent hydrogen, 25 per cent helium and about 2 per cent of the heavier elements; because of this, they are also gaseous in nature. The inner terrestrial planets of Mercury, Venus, Earth and Mars, did not form from sufficiently large perturbations to confine all the initial material, but resulted, in processes not fully understood, from chemical binding of the heavier elements to form rocky bodies from which most of the abundant elements, hydrogen and helium, escaped. The stage was set for the development of life on one of those terrestrial planets.

Speaking astronomically, the Sun is a very ordinary star and there are another thousand million like it in our galaxy alone, but to us it is the most important star in the whole Universe. It is the star that heats, provides and succours life on our own planet. This brings the writer to the central questions of this chapter: is there life anywhere else in the Solar System, to which a reasonably confident answer can be given of "no", and is there life anywhere else in the Universe, to which no confident answer is available at this time and only an outline of the possibilities and uncertainties can be given. The discussion is entirely an astronomical one; the question of the origin and evolution of life, one of the great questions facing the human race, is a biological one in which evolution of the present life species is clearly demonstrated but in which the problem of how life developed out of a world which originally had no life, still presents a formidable challenge.

The existence of life on Earth is clear to all, particularly its remarkable diversity and richness, which is constantly demonstrated by television programmes that portray so many forms of animal, marine, insect and plant life. The very close interrelation between different life species and their dependence on each other is also a matter of amazement. Life has evolved to match its environment exactly and has responded to the simple astronomical factors such as the Earth's rotation every 24 hours, which causes night and day, and the tilt of its axis, which gives the seasons as it orbits the Sun every year. Over the past 10 000 years or so, a rapid change in the evolution of life has been caused by the development of human societies which intervened in the evolution of animal and plant life. The abundant presence of cattle, sheep, pigs and poultry has happened entirely because they provide delicious and tender meat for the human race; their survival in the wild would have been very uncertain. Similarly, the extensive fields of corn, vegetables and orchards have been husbanded for human consumption. Of course, human control of life forms is only partial and is largely limited to the large mammals and plants. Microscopic forms of life, such as the insects and bacteria, still resist human attempts to control their evolution, although strenuous attempts are being made every day, particularly with the aim of controlling disease. But, in astronomical terms, the Earth's environment changes little over long periods of time, its orbit is very stable and the Sun's radiation is very constant over long periods of time. This makes it difficult to explain traumatic effects on life, such as the ice ages and the extinction of the dinosaurs, and can only mean that the balance of life on Earth is very delicate, like being on a knife edge, where small astronomical perturbations can have major effects on the development of life. It is also true that life itself can generate changes in its terrestrial environment by, for example, the generation of free oxygen into the atmosphere by trees and plants.

The possibility of life on other planets in the Solar System can be quickly narrowed to two – Venus and Mars. The giant planets are gaseous, are mainly composed of hydrogen and helium and, like Pluto, are far too cold, the warmest temperature lying at more than –100°C. Mercury, like the Moon, has no atmosphere and a surface temperature far above the boiling point of water at 300°C. This leaves Venus and Mars, and the possibility that life might exist on those planets has caused considerable excitement and received great publicity during the nineteenth and twentieth centuries. Venus was the planet whose overall parameters were closest to those of the Earth, being about the same size and mass and about three-quarters of the distance from the Sun as compared to the Earth. But it was shrouded in clouds which prevented its surface being observed from Earth. Mars was considerably smaller and lighter than the Earth and half as far away again from the Sun, but its atmosphere was transparent and its surface could be observed. Detailed observations were carried out by American and Italian astronomers towards the end of the nineteenth and into the beginning of the twentieth century; these reported geometrical lines on the Martian surface which were called canals, implying construction by intelligent life. These were not revealed in photographs of the planet, but were the result of visual observations where it was argued that the rapid time response of the eye could see such fine detail whereas the fluctuations of the Earth's atmosphere would blur them out in a photographic exposure. With the passage of time, these observations of canals have been shown to be spurious.

It was easy to imagine Venus, so like the Earth in mass and size and a little closer to the Sun, as a warmer and lush planet, possibly even more abundant with life as the Earth. But this exciting possibility quickly vanished with the post-war observations made with the early radio telescopes which, unlike the optical telescopes, could see through the cloud-covered planet to its surface. These revealed Venus to have a temperature of nearly 500°C; the hottest Earth-bound kitchen oven is less than 300°C and this result completely removed Venus as a possible abode of life of any kind. The radio studies also revealed another surprise, that is that Venus is rotating very slowly, with a period of 243 Earth days and in a retrograde direction, unlike other bodies in the Solar System; the reason for this retrograde rotation is not understood. The very high temperature of the Venetian surface was confirmed more directly by Soviet space probes which dropped landers on the surface in the early 1970s, and these transmitted measurements of the local temperature to Earth before the heat caused them to malfunction. The most detailed observations of the surface of Venus were made by the American space probe, *Magellan*, which orbited the planet in the early 1990s and mapped its surface with radar. These observations confirmed the high surface temperature and revealed major volcanic activity. (See Colour Plate 10.)

The search for life switched completely to Mars, to which American space probes, involving both orbiters and landers, were sent in the 1970s. The orbiters recorded the Martian surface in great detail, revealing giant volcanoes, the largest of which, named Olympus Mons, reaches an altitude higher than any mountain in the Solar System, more than three times the height of Mount Everest. Also seen were impact craters, and networks reminiscent of rivers and their tributaries, but now dry (see Colour Plate 11). Temperatures on the Martian surface are cold by Earth standards at more than 100°C below zero, but could rise to a comfortable 20°C above freezing point. The Martian atmosphere is devoid of free oxygen and composed mainly of carbon dioxide. Two of the American probes, Viking 1 and Viking 2, in addition to their orbiter, had a smaller spacecraft which was released to soft-land on the Martian surface, carrying instruments with the prime objective of detecting the presence of life, past or present. Images taken from each lander showed a completely barren surface devoid of any of the life forms familiar on Earth. Samples of Martian soil were scooped on board by an automatic shovel and subjected to chemical tests in a search for any metabolic activity of the form that would be caused by any life organisms on Earth. The results were negative; there was no evidence for any kind of biological systems on Mars, even of the most elementary kind, alive or dead. (See Colour Plate 12.)

After this chapter had been written, a development occurred which was given prominent worldwide publicity in the media. It was claimed that evidence had emerged clearly demonstrating that some form of primitive life had existed on Mars in the distant past. The reporting ignored the scientific method, which I have already strongly recommended to the reader, and I will attempt to apply that discipline now in discussing the matter. The evidence came from the analysis of the structure and detailed composition of a meteorite that was found in Antarctica and which was believed to have originated on Mars, largely because the abundance and isotopic composition of the trapped gases were the same as in the Martian atmosphere as measured by the Viking landers. The conclusion is a likely one but cannot be regarded as definitely proven. If correct, the only mechanism that could explain its arrival on Earth is of a massive impact on Mars by an asteroid or comet, which imparted sufficient energy to a piece of Martian rock to accelerate it to more than the velocity of 5 kilometres per second required to escape the planet. It then had to enter an orbit that crossed that of the Earth and ultimately led to a collision. Reliable methods of dating show that such an impact on Mars would have occurred 16 million years ago, and the meteorite would have struck Antarctica 13000 years ago. The analysis of the meteorite revealed the presence of a certain type of large molecule (polycyclic aromatic hydrocarbons) which can be formed by biological processes, but not exclusively. Also found were carbonate globules that are similar to some terrestrial bacterially induced precipitates.

To summarize, the evidence for some earlier primitive form of life on Mars is persuasive but far from conclusive. It will take many more scientific investigations to settle the issue, and these must be conducted within the framework of the scientific method and not be governed by a desire to hit the headlines: otherwise the truth will elude us.

The space experiments on Venus and Mars were carried out because they addressed one of the most basic questions of interest to the human race: is there life elsewhere than on Earth? The answer, as far as the Solar System is concerned, is no, with the possible exception of an early form of primitive life on Mars. This once again demonstrates that life on Earth is in a finely balanced state, since its two near and similar neighbours, Venus and Mars, one a little closer to the Sun, and the other a little farther away, are devoid of life. On Venus, the very high temperature is caused by a very strong greenhouse effect in which the incident solar radiation, which is greater than that reaching Earth, penetrates the clouds to reach the surface and thereby heat it; but its escape in the form of infrared radiation is effectively blocked by the Venetian atmosphere, which is composed almost entirely of carbon dioxide, a very strong absorber of the infrared. The effect is runaway in nature: an increase in temperature causes more carbon to combine with oxygen to form carbon dioxide, thereby increasing the temperature, which further increases the carbon dioxide, and so on. The end result is reached when essentially all the oxygen in the Venetian atmosphere has combined with carbon to form carbon dioxide, thereby halting the runaway process. In the Earth's atmosphere, most of the oxygen is in a free form because of its separation from carbon dioxide in the photosynthesis processes of vegetation; however, there is still a greenhouse effect, much to our advantage, as can be seen from the icy conditions on high mountains, which do not enjoy the full blanketing effect of the Earth's atmosphere. On Mars, the decreased solar radiation causes a runaway effect in the opposite direction. This glaciation process is one in which a decrease in temperature causes more ice to form; this increases the reflection of solar radiation into space, which in turn causes a further decrease in temperature, and so on until an equilibrium is reached. Hence, in the relatively small difference between the orbits of Venus and Mars, water is forced into a gaseous state (steam) in the former and into a solid state (ice) in the latter. Only on Earth is water in its liquid state, a requirement for life as we know it. But the finely balanced state of life on Earth cannot be assigned to astronomical conditions only; life developed very slowly in a hostile environment and finally broke through to create and control a more friendly environment with an atmosphere carrying an abundance of free oxygen.

The question as to whether life exists anywhere else in the Universe other than on our own planet is one that has fascinated human beings ever since the dawn of serious intellectual inquiry. Thus, in 400 BC the ancient Greek philosopher, Metrodonis, wrote "It is unnatural in a large field to have only one shaft of wheat and in the infinite Universe only one living world". In 50 BC the Roman poet–philosopher, Lucretius, wrote "Nothing in the Universe is unique and alone, and therefore in other regions there must be other earths inhabited by different tribes of men and breeds of beast". During the Renaissance the Italian monk Giordano Bruno went further and insisted that "there must be an infinite number of suns with planets with life around them". However, as recounted in the story of Galileo's trial, his burning at the stake in 1600 was almost certainly attributable to his espousing the Copernican theory than to his imaginative view on extraterrestrial life. In 1690 the Dutch physicist, Christiaan Huygens, argued in favour of life on many other planets around many other suns, but used phraseology that seemed designed to make it difficult for the Church to take offence: "Barren planets, deprived of living creatures that can speak most eloquently of their Divine Creator, are unreasonable, wasteful and uncharacteristic of God, who has a purpose for everything". The next point in this potted history came in 1830 when the German mathematician, Carl Friedrich Gauss, suggested that an attempt be made to inform any intelligent life forms that may exist on other planets of the existence of intelligent life on Earth. He proposed the construction of a giant forest in the shape of a right-angled triangle with squares on each side. He argued that this demonstration of a knowledge of the famous geometrical theorem of Pythagoras would convince any extraterrestrial observer of the presence of intelligent life on Earth. Since the surface of the cloud-covered Venus cannot be observed from Earth, then, by the same token, the Earth cannot be observed from the Venetian surface. Hence, the closest possible extraterrestrial candidate to conduct the Gauss test had to be Mars, given that the barren moon could be eliminated. But, as Gauss could easily have calculated himself, detection of the shape of any gigantic Pythagorean forest was well beyond the capability of any Martian telescope, even of the largest size available today. This is true even if the telescope is figured perfectly and pointed perfectly, because there is an absolute limit to the resolution capability which is imposed by the wave nature of light and is called the *diffraction limit*. An approaching alien spacecraft would have to get closer than the Moon before it could detect the signs of intelligent life, such as the patchwork of cultivated fields and motorways, on Earth. Nevertheless, the nineteenth-century proposal by Gauss was the first ever to suggest a possible communication with intelligent extraterrestrial life, and this has become, in a much more highly sophisticated form, an accepted but controversial activity today.

Having established that the Earth is very likely to be the only planet in the Solar System to harbour life, the wider question can now be addressed as to whether life exists anywhere else in the Universe. The most reasonable starting point, perhaps the only starting point, is the premise that the only likely places are planets around other stars. Indeed, because of the unique presence of life on only one planet in the Solar System, this statement can sensibly be narrowed further to saying that the most likely locations of life are Earth-like planets revolving around Sun-like stars, but, of course, there could be other forms of life, different from our own, that develop on other kinds of planet orbiting other kinds of star. However, the chances of any *direct* observation of such life is zero with present-day technology and will certainly remain so for the foreseeable future. As we have already seen, it would take an approach to interlunar distances before life on Earth could be detected, and other planetary systems are at immensely greater distances, so great that even a direct observation of another star's planets is a very difficult problem. The diffraction limit, referred to above, cannot be reached by Earth-bound telescopes because of the shimmer of the atmosphere, but the Hubble space telescope orbits above the atmosphere and can now, after its repair, operate at its diffraction limit of about one tenth of a second of arc. At the distance of the nearest stars (about ten light years) this translates into a resolution distance of about 50 million kilometres, that is, two objects would just be resolved at ten light years if they were separated by 50 million kilometres and were of the same brightness. Although the Earth–Sun distance, at 150 million kilometres, is three times that, the brightness of the Earth is negligible compared to the Sun and would not be seen. It is clear that, at present, the chances of directly detecting Earth-like planets around even nearby stars are remote, never mind the possible presence of life on them. This situation could change in the future. NASA and other space agencies have been considering proposals for space-based interferometers, comprising a few telescopes separated by distances that would give them a resolution of more than 10 million times that of the Hubble space telescope, and therefore the capability of seeing planets around nearby stars. Also, if the spectra of such planets (seen by reflection from their parent star or by direct emission in the infrared) revealed the presence of free oxygen (O_2), this would afford very strong evidence for the existence of plant life, which is the source of free oxygen in our own atmosphere.

Of course, there are other ways of inferring the presence of planets or planetary material around stars. One is the classic method of using the Doppler effect, in which the motion of planets will cause a slight wobble in the central star, but this will be dominated by the heaviest planets and, if the Solar System is our guideline, Earth-like planets will be undetectable. Another method is the use of infrared observations where planetary emissions are no longer swamped by the parent star. Such observations have been made by the Infrared Astronomical Satellite (IRAS), which has detected infrared emission extending from some stars in the form of disks. These must be caused by rings of particles, like those orbiting Saturn, and not planets, since these, like the solar planets, would not be visible by this technique.

The possibility of detecting the presence of life on planets orbiting other stars therefore appears to be somewhat remote, but there is one other possible way of detecting extraterrestrial life – if that life is intelligent, if it wishes to communicate with other extraterrestrial civilizations, and if it has developed the high technology needed to do so. These conditions are generally met by the human race, so a good starting point is to ask how this could best be done if we attempted it from Earth? The answer is certainly not with a giant triangular forest with squares on its sides, as proposed by Gauss, since this would not be resolvable from the nearest planet, never mind the nearest star, and, in any case, light emission from the Earth would be completely swamped by the brightness of the Sun. For the same reason that we communicate by radio waves over large distances on Earth and to satellites and space probes, plus the fact that the Sun is very faint at such radio frequencies, we would attempt any interstellar communication by radio transmission. But which frequencies would we choose to transmit any interstellar message? Since the most likely detection of any signal would be by radio astronomers who may be observing our Sun (although it is not the most astronomically exciting object in the galaxy), then frequencies of astronomical significance should be selected in order to maximize the chance that the extraterrestrial radio telescopes are tuned into them. One such frequency is a fundamental transition in atomic hydrogen, the most abundant element in the Universe; from quantum mechanics (see Ch. 7), its ground state has only two levels, one in which the spin of the electron is parallel to that of the nucleus (a proton) and the other in which it is anti-parallel. When the electron flips from one mode to the other, the resulting emission (or absorption) has a frequency that gives a wavelength of 21 centimetres. Having selected a frequency for transmission, the next stage would be to use the language of the computer, and transmit a sequence of binary digits (bits) which could easily be deciphered and transformed into simple patterns, geometric forms and information about the sender. Such a transmission could be generated by using a radio telescope in an opposite mode of operation, that is by generating the signal at its receiver point which would then strike the telescope dish and be beamed out into space. The final decision would be to select the region of sky towards which such a beam should be directed. Since the only intelligent life that we are aware of is ourselves, the best likely candidates are stars like the Sun, and those which are nearest to the Earth should be selected so as to maximize the signal and to reduce the time delay which at best has to be several years.

On the chance that some advanced extraterrestrial civilization had followed the same logic as outlined above, and had beamed radio signals containing intelligent information towards the Solar System as a candidate for carrying a life-bearing planet, several radio telescopes devoted some time to observing nearby (within about ten light years) solar-like stars in a region embracing the wavelength of the hydrogen line at 21 centimetres. No signal has been detected that indicated the possibility of an intelligent message transmitted by civilizations on planets orbiting the stars selected.

The inverse process has also been carried out, but this should be regarded as more of a public relations exercise than an experiment. The largest radio telescope in the world was built by placing a 300 metre diameter mesh over a naturally shaped concave valley in Arecibo, Puerto Rico. A major upgrade of the telescope in 1974 led to a re-dedication ceremony in which a powerful signal was directed towards the globular cluster M13 on 21 centimetre wavelength. The beaming and single frequency nature of the transmission was such that it would be ten million times more powerful at 21 centimetres than the Sun (but still weak) when it reached M13. The transmission lasted 3 minutes, contained 1700 characters of two types to indicate binary digits (bits) in a simple code which any advanced intelligence would have little difficulty in cracking. The information included atomic numbers and weights of the elements, and details of chemical compounds, the DNA helix and the form of the Solar System. M13 was selected because it was within the limited beam direction of the Arecibo telescope and because it contains 300000 stars, thereby increasing the chance that one might have an Earth-like planet with an advanced technological society. But globular clusters were formed in the very early stages of the Universe and the galaxy and therefore contain very few of the elements necessary for life as we know it. Also, in case the reader is waiting with baited breath for the conclusion of this exercise, the writer has to say that the distance to the M13 globular cluster is 24000 light years, so the earliest possible reply is 480 centuries away.

These attempts to initiate some form of interstellar communication between intelligent and technologically advanced societies did not demand many resources and were often carried out in a casual or even flippant manner. But they did trigger a wider debate on the whole question of communication with extraterrestrial societies which established the matter as a wide political issue rather than a narrow scientific one. The reason is that the human race has acquired the technology for radio inter-stellar communication only over the past 50 years. This is a tiny fraction of the time taken for the development and evolution of life, and means that any extraterrestrial civilization we may contact will almost certainly be far more advanced than ours. Past experience on Earth shows that, in the interaction between different societies, the less advanced have always suffered in ways varying from subjugation to actual extinction. Of course, we are separated by at least several light years from the nearest, and pos-sibly hostile, extraterrestrial civilization, and travel over such distance, certainly in human terms, seems unimaginable until the far future. Never-theless, it seems prudent not to advertise our presence, and to adopt only a listening mode rather than a transmission mode, and this indeed is what has happened.

The individual and sometimes ad hoc attempts to establish interstellar radio communication were replaced in the early 1990s by a coordinated programme, funded by the American space agency NASA in a systematic radio monitoring of the skies to Search for Extra Terrestrial Intelligence (SETI). This built special sophisticated receivers, designed for the pur-pose, and time on the world's most powerful radio telescopes was rented to conduct a systematic programme, both in a survey of the sky and in its range of frequencies. It was formally launched in 1992 to mark the 500th anniversary of the voyage of Columbus to the New World, but was can-celled by the US Congress two years later. Attempts to continue the pro-gramme with alternative funding are still being made, but so far it has not detected any signal which indicated any kind of intelligent source.

It is now appropriate to attempt an assessment of the likelihood of life existing outside our own Solar System. The reader should understand that there is no substantive basis for such a calculation, because our entire knowledge of life is based on its existence in only one place in the Universe – the Earth. The assessment will be based on the very simple assumption that any possible life will need a planet of the same environ-mental conditions as existing on Earth. This means a planet of the same size as the Earth, with the same composition as the Earth and orbiting at about the same distance as the Earth is from a star like the Sun. Whether other forms of life of a totally different nature to those on Earth can exist in conditions totally alien to those we experience is beyond the scope of this book and the experience of the writer. It should be stressed that, even with this very simple approach, any estimate is extremely uncertain, as will become evident. The starting point is the easiest and the most certain; we do have a good idea of the number of stars in our galaxy which are similar to the Sun. More than half of these are multiple, usually having two stars, and could also have planets on which life could evolve but which would presumably have major differences because of the presence of two suns. If only the single stars are taken and it is assumed that they all have planetary systems (a not unreasonable assumption), then there are a few thousand million planetary systems around Sun-like stars in our galaxy. At this point the major uncertainties are reached: how many of these planetary systems would contain a planet like the Earth? Bearing in mind the knife-edge situation of the Earth as far as life is concerned – even the similar and nearby planets of Venus and Mars are devoid of life – there is no way of answering this question, even statistically, with any meaningful precision. So, here I only offer a numerical example in order to give the reader some feeling for the numbers involved. If one in every thousand Sun-like stars has a planetary system that includes an Earth-like planet, then there would be a few million "earths" in our galaxy, of which only a few would be within a hundred light years. If the next big step is taken by assuming that an Earth-like environment will lead to the development of life, for which the *only* basis is the existence of life on Earth, then one would conclude that life probably exists somewhere else in the galaxy. But, as has been explained, there is no possibility of directly detecting such life, and the only possibility is that an intelligent, highly technological, extraterrestrial civilization decides to transmit radio signals announcing its existence. The possible detection of extraterrestrial life is therefore confronted by yet another major uncertainty: the possibility of any life on another Earth-like planet having evolved to an intelligent stage, developed an advanced technology and, unlike us, decided to beam out radio waves.

A fair conclusion to the question of life elsewhere in the Universe is that it is more a matter of belief than of calculation. I follow a Copernican philosophy, which cannot accept the concept that the Earth is in some way unique and, therefore, I believe that there is other life on other worlds beyond the Solar System, probably in the galaxy, but almost certainly somewhere else in the Universe, where the immense number of possible locations dwarfs even the large uncertainties in our present calculations. But whether extraterrestrial life has developed to a sufficiently high technological state, and within a sufficiently close distance, say less than a hundred light years, in order to effect meaningful interstellar communication, does seem extremely remote. But, like the other extremely remote possibility of an asteroid collision, it would mark, for totally different reasons, a milestone in human history.

CHAPTER ELEVEN
The Stars –
their Birth, Life and Death

The Stars with deep amaze
Stand fixt in stedfast gaze,

. . .

But in their glimmering Orbs did glow,
Until their Lord himself bespoke, and bid them go
Milton

The stars are the most important building blocks of the visible Universe and their nature and evolution is therefore one of the fundamental problems in astronomy. The first steps in understanding their physical nature were made in the 1920s, notably by Arthur Eddington, who represented them as large gaseous bodies in hydrostatic equilibrium in which the weight of the star was borne by the high pressure in the intensely hot interior. This allowed the temperature and density distribution to be determined from the centre to the surface of the star in order to explain the observed emerging flux of radiation. As has already been related (in Ch. 8) a major step forward was taken in the 1930s with the realization that the energy source of the Sun, and presumably therefore of all other stars, was the thermonuclear conversion of hydrogen into helium. But what happens when the hydrogen is exhausted and why is there such a wide variety of stars in the sky? These questions are basic to the problem of stellar evolution and were tackled in the post-war years when major progress was made to the point where we can now provide reasonable, but not complete, answers.

Stars are born from gravitational contraction in the interstellar medium, which is composed mainly of gas with a small dust component. The gas is mainly hydrogen with some helium and a little of the heavier elements; the composition of the dust is not known precisely but certainly contains some carbon and silicon. The total mass of the interstellar medium is only a few per cent of that of stars, but stars are still being formed from it. This is known because, as will shortly be shown, the most luminous stars can only have brief lifetimes, in astronomical terms, of about a few million years when they will have used up their hydrogen fuel. Less is known about the actual birth of stars than the subsequent stages of their evolution, but the accepted picture is of a gravitational collapse in the interstellar medium induced by some local perturbation. Dense regions of the interstellar medium exist which are considered to be areas of star formation, but are impenetrable to visible light and can only be observable with high-frequency radio waves. Details of the process of star formation, and whether more than one star or planets are formed are not fully understood, as was explained in the discussion of the formation of the Solar System. (See Colour Plate 13.)

Once a contraction of sufficient material in the interstellar medium has occurred, subsequent events are much better understood. The release of gravitational energy heats the contracting matter and if the temperature in its core becomes high enough (about 10 million degrees Kelvin), thermonuclear burning of hydrogen into helium will begin and we have a star. This is a very finely tuned thermonuclear generator, which is regulated by gravity; the energy generated maintains the high temperature in the interior and the resulting kinetic pressure and radiation field is gravitationally balanced by the weight of the star. The thermonuclear fusion process is a sensitive function of temperature, so even a small rise in temperature will result in a significant increase in energy generation which, if it caused a further increase in temperature, would produce more energy again and an explosive situation would be reached; but the increase in temperature increases the pressure, which lifts the star and causes it to expand slightly and therefore cool slightly, thereby preventing an internal temperature rise. Conversely, a small drop in the temperature of the core will reduce its rate of energy generation, but the consequent reduced pressure will cause the star to contract, thereby heating it and maintaining its original temperature. In other words, the fluctuations in thermonuclear energy generation caused by variations in temperature are accommodated by the gravity of the star, which absorbs increases and replaces decreases, and therefore maintains a perfectly balanced system.

The nature of the star that is formed and, indeed, its future evolution depends almost entirely on one parameter, its mass (i.e. the amount of material that is gravitationally captured in its collapse). The chemical composition also plays a role, but this is minor because it does not vary greatly, being dominated by hydrogen and helium, which comprise 98 per cent or more of the mass. This means that stars of the same mass will have similar characteristics just after their formation and will evolve at the same rate and in the same way. The range in stellar masses is not exceptionally large, varying from the lightest at one-tenth of a solar mass to the heaviest at about 50 to 60 times a solar mass. The lower limit of a tenth of the mass of the Sun is set by the fact that lighter contractions do not have enough gravitational energy to heat the interior to a sufficient temperature to trigger a thermonuclear reaction; smaller condensations can form planets, as in our Solar System, or larger objects which have just failed to ignite; these rely only on the gravitational heating of contraction and are known as brown dwarfs. The upper limit of 50 to 60 solar masses is set by the fact that heavier condensations produce such a high interior temperature that the pressure of the material and the radiation is enough to overcome the gravitational force and prevent contraction. The least massive stars are red in colour, with a surface temperature of about 3000° and have a luminosity (total rate of energy emission) which is only one ten thousandth that of the Sun. The most massive stars are blue in colour, with a surface temperature of about 50000° and have a luminosity 10000 times greater than the Sun. There is therefore a sequence of stars, known as the main sequence, which is composed of those in their initial state of hydrogen-burning (as most stars in the sky still are) and which is determined completely by the masses of the stars.

It is clear that the massive stars are emitting radiation at a far higher rate than the lighter stars, even compared to their greater energy reserves. Consequently, they will use up those reserves more quickly and have much shorter lifetimes. Since the thermonuclear burning of hydrogen into helium represents the star's energy reservoir, and since that process releases about 1 per cent of mc^2, a simple, but very rough calculation can be made of the time needed to use up that reservoir at the observed luminosities. As will be revealed shortly, it happens that a star will change significantly and leave the main sequence when the hydrogen in its core, about 10 per cent, has been exhausted, and the following calculations have been based on that premise. The lifetimes of the hottest, most massive stars are about one million years, of the Sun about ten thousand million years and of the coolest, least massive stars about a million million years. For comparison, the age of the Milky Way galaxy is about 12 to 15 thousand million years and so most stars are still burning hydrogen in their cores and lying on the main sequence. However, the hottest stars must have been formed recently in astronomical terms since their lifetimes are much shorter than the age of the galaxy, and any heavy stars formed in earlier epochs will have now completed their whole life-spans in a way to be described later.

Before describing the evolution of stars through the whole of their lifetimes, it is appropriate to revise and extend the nuclear physics developments discussed in Chapter 7. This is because the evolution of stars is governed by the nuclear processes that occur in their interiors, a fact that was first realized by Fred Hoyle in the immediate post-war years and which led him to lay the foundations of the subject and, in parallel, to launch an attack on the great question of the origin of the elements. Drawing on brilliant collaborators in the persons of Margaret Burbidge, Geoffrey Burbidge, and Willy Fowler, a seminal paper was written in 1957 which formed the basis of future studies in stellar evolution. There is little doubt that the leading figure in this exercise was Hoyle, and therefore it is a great pity (some would say an injustice) that the recognition of its value, by the award of a Nobel prize, fell on Fowler to the exclusion of Hoyle.

Thermonuclear fusion of the positively charged nuclei in stellar interiors is achieved only by the most energetic or fastest-moving particles, because they have to overcome the strong electrical repulsive forces between them in order to reach the energy pit of the strong nuclear force. As was explained earlier, the particle does not have to surmount the repulsive barrier completely because its wave nature allows it to tunnel through near the peak. The thermonuclear fusion of four hydrogen nuclei into one helium nucleus, the main energy source of the stars, does not happen in a single binding process, because the chance of four protons colliding simultaneously with sufficient energies to overcome the combined repulsive barrier is negligible; the fusion process has to go in steps of binary collisions. It starts with the collision of two protons which fuse to form the heavy hydrogen isotope, deuterium, consisting of a proton and neutron, with the emission of a positron and a neutrino. The positron will quickly find its antiparticle, the electron, and they will annihilate with their masses being converted to pure energy in the form of gamma rays; the non-interactive neutrino will pass through and escape from the Sun. The deuterium nucleus will ultimately collide with another sufficiently energetic proton to form the light isotope of helium, composed of two protons and one neutron, with the emission of energy in the form of a gamma ray. That light helium nucleus can then fuse with a similar one produced by the same process to produce the normal helium nucleus (an alpha particle) composed of two protons and two neutrons; the conservation of charge and mass results in two protons also being released and these can continue in the fusion processes. There is another source of energy generation by the fusion of hydrogen into helium, in which carbon and nitrogen act as catalysts, but the end product is the same: four protons ultimately become one helium nucleus, with the release of two neutrinos and the generation of a remarkable 26 million electron volts of energy.[1]

An important question concerning the later stages of stellar evolution

---

1. An electron volt is the energy acquired by an electron when accelerated through an electric potential of 1 volt (see glossary).

Plate 1 This painting on papyrus, using the ancient method of laying the leaves in both directions and leaving them to dry naturally in the Sun, shows the zodiac of Dendara that decorated the ceiling of a Ptolemaic temple in the major town of Dendara in ancient Egypt. The original is situated in the Louvre Museum in Paris. In addition to the signs of the zodiac, the papyrus also shows: *sa'mut*, the hippopotamus with a crocodile on her back; *Meskhetiu*, the bull; and *Selket*, the scorpion goddess. This clearly demonstrates the interlinking of astronomy and mythology in ancient times.

Plate 2 A long time exposure shows the star trails above the Anglo-Australian telescope dome on Siding Spring Mountain in Australia. The effect clearly demonstrates why the early astronomers believed that the stars were **fixed** on rotating spheres.

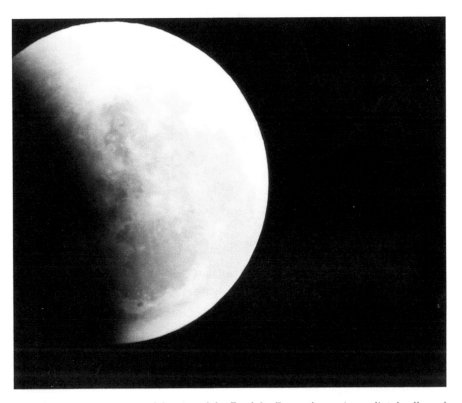

Plate 3 The measurement of the size of the Earth by Eratosthenes immediately allowed Aristarchus to determine the distance to the Moon by using a partial lunar eclipse. This allowed the size of the Moon to be measured compared with the Earth, and from its angular diameter (about half a degree) its distance could be estimated. This was the first step on a long, long road to measuring the distance of astronomical bodies.

Plate 4 The first reflecting telescope, designed and built by Isaac Newton, which still exists in the rooms of The Royal Society in London. Also shown is the title page of his treatise on *Opticks*. The light enters the tube on the right-hand side and strikes a concave mirror at the base, from which it is reflected upwards in a converging beam to strike a plane, inclined mirror which reflects it out sideways where it can be viewed by the eyepiece shown. All major modern telescopes are reflecting in nature but the flat, inclined mirror is replaced by a curved, convex mirror which reflects the light back down the tube and through a hole in the centre of the primary mirror where it can be viewed or processed.

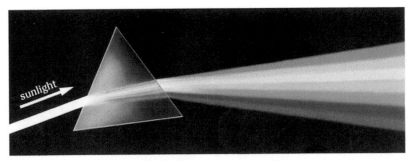

Plate 5 Isaac Newton's separation of sunlight into its different colours by means of a prism was the starting point for the later development of astronomical spectroscopy or astrophysics.

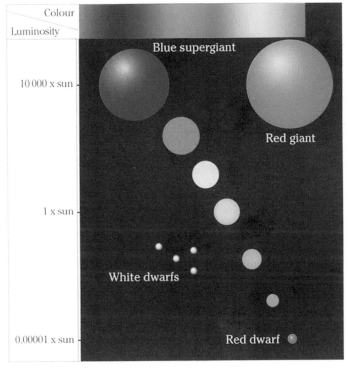

Plate 6 This illustration shows the wide variety of stars in terms of their colour (temperature) and luminosity (total energy output). Most stars lie in a sequence which extends from the hottest blue supergiants down to the faintest red dwarfs, which also marks a range in mass from fifty times the mass of the Sun to a tenth of that of the Sun. In addition to the stars of this main sequence, there are red giants and white dwarfs, which are also shown. The reason for this distribution of stars and how they evolve from their present states is a basic problem in astronomy and is addressed in Chapter 11.

(a)　　　　　　　　　　　　　(b)

(c)　　　　　　　　　　　　　(d)

Plate 7 A selection of galaxies. (a) A typical spiral galaxy (NGC 29.97) showing its nucleus and spiral arms. The nucleus is yellow because of the high concentration of old stars, whereas the arms are blue because of the young stars recently formed from the interstellar medium. The interstellar dust shown by the dark lanes surrounding the nucleus and the interstellar gas is revealed by the pink spots where the hydrogen is glowing in the ultraviolet radiation fields of the hot stars. (b) Another spiral (M83), which is considered to be very similar to our Milky Way galaxy. It is at a distance of 27 million light years. (c) A giant elliptical galaxy (M87) which is at the heart of the Virgo cluster and has no interstellar medium left and therefore no young blue stars. It has several thousand globular clusters, many of which can be seen just beyond the edge of the central part. (d) The large Magellanic Cloud, an irregular galaxy which is a companion to our own Milky Way galaxy. It is blue because hot stars are still forming from its rich interstellar medium and their ultraviolet radiation excites the hydrogen gas causing the pink glow, which is clearly seen in the left-hand nebulosity called 30 Doradus (where the 1987 supernova explosion occurred).

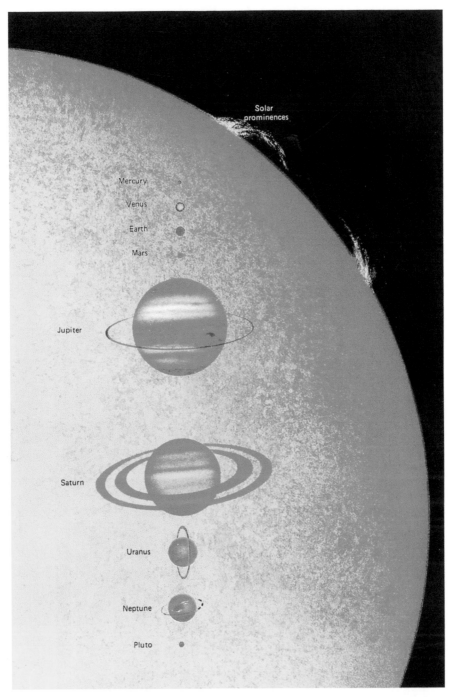

Plate 8 The size, nature and colours of the planets shown to scale against the background of an image of the Sun.

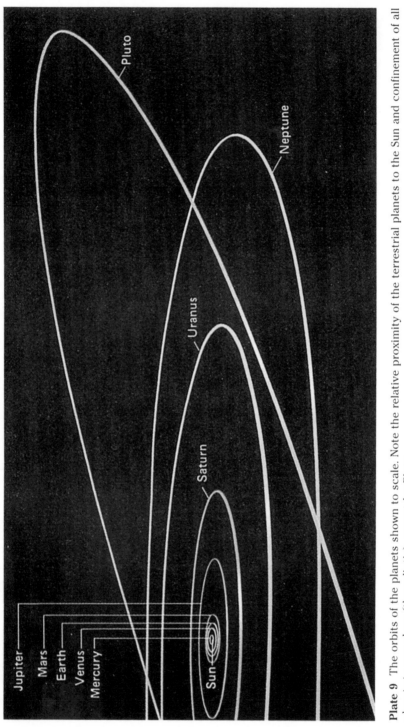

Plate 9 The orbits of the planets shown to scale. Note the relative proximity of the terrestrial planets to the Sun and confinement of all planets to one plane (the ecliptic), except for Pluto.

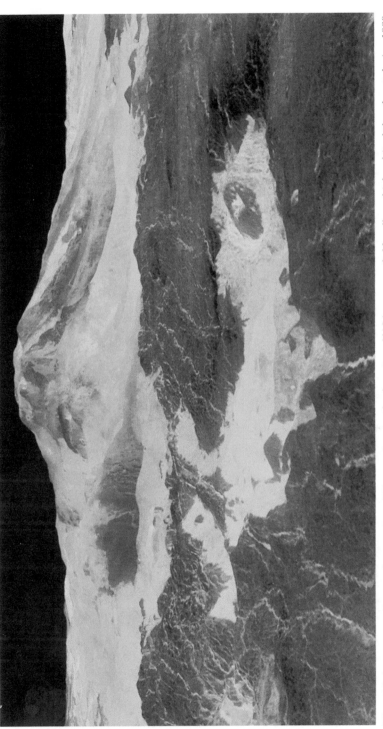

Plate 10 A composite image of one region of Venus taken by the American NASA spacecraft *Magellan* which orbited the planet during 1989-90 and mapped the surface using radar thereby penetrating the dense atmosphere. The image is a computer simulation of how the surface would look to the human eye and shows a large volcano (named *Maat Mons*) and its extensive lava flows. The typical surface temperature beyond these flows is 500°C.

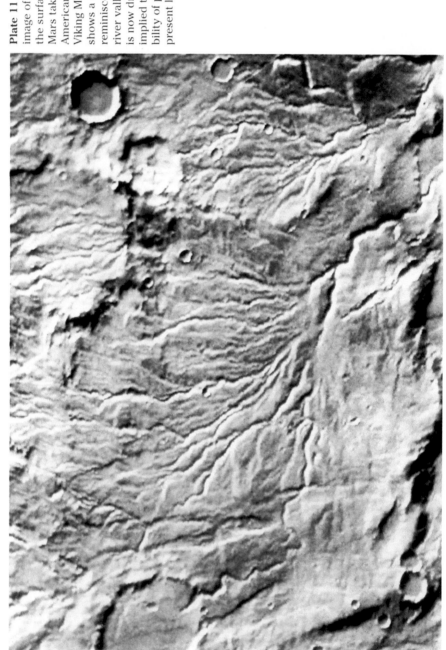

Plate 11 An image of part of the surface of Mars taken by the American NASA Viking Mission. It shows a complex reminiscent of a river valley, which is now dry. It implied the possibility of past or present life.

Plate 12 An image taken with the Viking lander sitting on the surface of Mars showing the dry, brown terrain. Scoops of soil were taken into the lander and analysed chemically but revealed no sign of any form of biological molecule, living or dead.

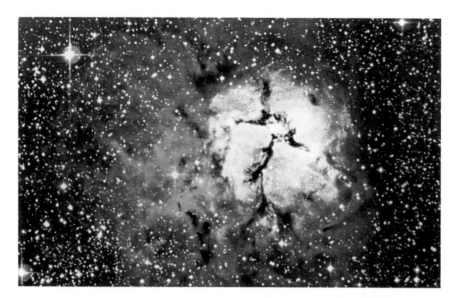

Plate 13 The *Trifid Nebula* is a region of recent star formation in which the massive, young, hot stars excite the remnant hydrogen gas from which they formed to produce the characteristic pink glow. The dust from the remnant cloud is also revealed by the dark lanes in the pink emission nebula and by the blue reflection nebula in which the dust grains are reflecting the blue starlight.

Plate 14 Known as the *Helix Nebula*, this is the nearest planetary nebula, being only 400 light years from the Earth and its apparent size is about the same as the Moon. It represents a solar-type star in the later stages of its evolution, having shed its atmosphere in the red giant stage: this is seen as the circular nebulosity which is glowing in the radiation field of the white dwarf into which the stellar core has collapsed.

Plate 15 The 1987 supernova explosion in the 30 Doradus region of the Large Magellanic Cloud which, at 170 000 light years, is the nearest galaxy to our own. On the left is a pre-supernova image with the progenitor star arrowed, and the image after the event has occurred is shown on the right.

Plate 16 The open star cluster called the Pleiades which is 400 light years away. It is also called the Seven Sisters because of the brightest stars which can be seen with the naked eye. It contains 120 stars and is a young cluster of age 50 million years. The massive, young, hot stars are clear from their blueness and the dust particles in the remnant interstellar medium are revealed by their reflection of the blue starlight.

Plate 17 A globular cluster in the Large Magellanic Cloud which contains several thousands of stars. Similar clusters surround the Milky Way and were formed in the very initial phase of the evolution of our galaxy. Their ages place constraints on the age of the Milky Way and of the Universe. The oldest have ages which approach 15 thousand million years.

Plate 18 An image of the quasar 3C 273 in green light showing its jet in the right-hand corner (marked by the square). The general jet velocity is towards us and there must be another exactly opposite, which is not seen because the extremely high receding velocity Doppler-shifts the light out of the visible region and into the infrared.

Plate 19 A Seyfert galaxy in the southern constellation of Doradus some 50 million light years away. Its central, bright nucleus is evidence of a black hole which is fed by surrounding material pouring into it like a quasar, but the nucleus is not dominant as in a quasar and the surrounding galaxy and spiral arms are clearly visible. Astronomers are of the opinion that every spiral galaxy has a black hole in its nucleus, but the normal systems like our own Milky Way have largely used up the material fuel which feeds it.

is whether fusion of the light elements beyond helium is a possibility. Readers will remember from the discussion of nuclear physics in Chapter 7 that energy is available from the fusion of the light elements all the way to iron, which is the most closely bound nucleus. Most of that energy, about 80 per cent, is released in the fusion of hydrogen into helium, which can be achieved with temperatures of about 10 million degrees Kelvin. Clearly, further fusion to heavier elements with their higher nuclear charges and therefore greater repulsive forces, will require much higher temperatures, and the knowledge of nuclear physics at the time indicated that the jump from helium to carbon was so great as to severely limit it happening in stellar interiors. But carbon was present in the stars, it was present on Earth, and the astronomical evidence that it was not present in the primeval material of the Universe led Hoyle to propose a nuclear resonance in carbon that would facilitate its fusion into carbon. This triple alpha process, in which beryllium is an intermediary, allows three helium nuclei to fuse into a single carbon nucleus, with the release of more than 7 million electron volts of energy, and is initiated at a temperature of about 100 million degrees Kelvin. Further fusion to heavier elements would require even higher temperatures to overcome the higher repulsive electrical forces between the more highly charged nuclei. To burn carbon all the way to the ultimate element, iron, beyond which no further nuclear energy is available by fusion, requires temperatures as high as a thousand million degrees. These three temperatures – 10 million degrees to initiate the fusion of hydrogen into helium, 100 million degrees to allow the formation of carbon from helium, and 1000 million degrees to take carbon through all the elements up to iron – are three critically important parameters which decide, as will be seen, the routes a star will take in the course of its evolution. Another question of great interest is the origin of the elements which are heavier than iron and which are present in the stars and on the Earth. These could not be formed by the thermonuclear processes described above and, if they were not present in the primordial material, some substantial energy input was required to build them up from the lighter elements. This is another issue which is addressed in the following account of the evolution of stars.

There are different routes of stellar evolution which have very different characteristics and which also vary markedly in their timescales. The route followed depends entirely on the initial mass of the star and is different for low-mass stars, for high-mass stars, and for very high-mass stars. The dividing points in mass are not known precisely, but are about eight times and twenty times greater than that of the Sun. Therefore, the

Sun classifies as a low-mass star and its evolution will now be discussed as a representative example.

A model of the Sun can be set up from knowing its size, mass and luminosity. It is assumed to be in hydrostatic equilibrium in which the gas pressure and radiation pressure balances the Sun's gravity. This leads to a determination of the distribution of temperature and density throughout the interior. The temperature of the core turns out to be 15 million degrees, sufficient to initiate a thermonuclear reaction and allowing the rate of that reaction to be calculated. It is only in the core that hydrogen is burned into helium; as the temperature drops from the central 15 million degrees Kelvin to below 10 million in the outer regions, thermonuclear processes cease and the temperature continues to drop until the value of about 6000° is reached at the photosphere. The intermediate layers are heated by the outflowing radiation flux, which is the main cause of energy loss, but since the upper regions cannot carry such a high radiative energy flow, turbulence is induced; however, the resulting energy flux carried by convection is very small compared to radiation and it is therefore neglected in the calculations, and only radiative energy flow is treated as the means by which the Sun cools.

The Sun is currently in this phase, sitting on the main sequence, in which it is burning hydrogen into helium in its core. It is a very stable phase in which the energy generation is finely regulated by the Sun's gravity, and during which its energy output will vary only very slightly. It represents the longest period of the Sun's lifetime and does not end until all the hydrogen in its core is exhausted and transformed into helium; this will take a total of 10000 million years and since the Sun is currently 5000 million years old, it is half way through its present stage and has another 5000 million years to go.

When the hydrogen in the Sun's core is finally exhausted, representing about 10 per cent of its mass, the Sun will enter a relatively rapid period of change. The remnant helium core is no longer providing energy and it contracts and is therefore heated by the Sun's gravity until the temperature in the shell immediately surrounding the core, which is still hydrogen-rich, is increased to a level that initiates further hydrogen-burning. At this stage there is no energy generation in the core but only in a shell surrounding it, but the continued heating of the inner helium core by the hydrogen-burning shell finally raises its temperature to the 100 million degrees needed for helium-burning, which then proceeds via the triple alpha process described above. Energy is now being generated in the inner core by the burning of helium into carbon, and in the immediately surrounding shell by burning hydrogen into helium. These reactions continue until the helium in the inner core is exhausted and its composition is mainly carbon. It is surrounded by a helium shell, whose hydrogen has been exhausted and which has been heated sufficiently to start the burning of helium into carbon; beyond it, in an outer shell, hydrogen is being burned into helium. At this point the progression of nuclear reactions, in which successively heavier elements are heated sufficiently to start thermonuclear burning, ceases. In the Sun, the ultimate carbon core will never reach the ultra-high temperature of 1000 million degrees needed to burn it to the heavier elements all the way up to iron.

In the development of these stages of thermonuclear reactions in the Sun, a severe hiccup will occur on the initiation of helium-burning in the hydrogen exhausted core. This is because an important physical change will have occurred in the core before helium starts to burn: it will have been compressed to such a high density as to become degenerate. This is a state of matter that can only be treated by quantum physics and will be explained in those terms when the ultimate stage of the Sun's life is described, because that is also a degenerate but stable state. You are asked to bear with me until then and accept at this point the fact that, in a degenerate gas, the pressure depends only on its density, unlike a normal gas where it also increases with temperature. The consequence is that the increase in temperature caused by the energy generated by the onset of helium-burning causes a further increase in energy generation but no increase in pressure. The normal control mechanism of a star, in which an increase in temperature would cause an increase in pressure, with a consequent expansion and cooling, is no longer operative and the core enters an explosive condition. However, a total explosion is averted because, when the core temperature exceeds about 300 million degrees, the electrons have sufficient energy to break the degeneracy condition, the pressure once again increases with temperature, and the helium core then burns in a controlled fashion with the normal regulative process in operation. This effect is known as the *helium flash*, because it is very rapid in astronomical terms, lasting only about one year.

The Sun as a whole will start to change markedly and rapidly as soon as its inner core is exhausted of its hydrogen fuel. The burning of hydrogen in a shell surrounding the core, and the subsequent burning of helium in the core, will cause the Sun to expand and it will continue to do this, except for the blip caused by the helium flash, until it becomes a red giant with a surface temperature of about 3000° Kelvin. Although cooler than the Sun in its present state, its size is so great that its total emission is several thousand times greater. At this point, the Sun will have expanded to a size that embraces the Earth's orbit, thereby ending any life on our planet if, indeed, any life on Earth has survived at that time; but there are still 5000 million years to go!

The Sun will reach its red giant stage 100 million years after it has burnt its hydrogen core, a relatively short time compared to its lifetime in its

211

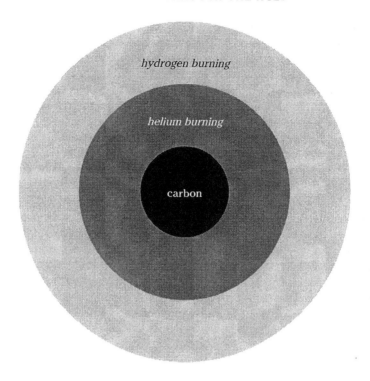

The core of a solar-type star, which has reached its red giant stage. The inner core has had its hydrogen burned first into helium, which was then burned into carbon at a temperature of 100 million degrees Kelvin. But the mass of the star is not great enough to induce the ultra-high temperature of 1000 million degrees needed to burn carbon into heavier elements. Hence, no more energy is generated in the inner carbon core, although some is still being generated in the surrounding helium shell, some of which is being burned into carbon, and in the outer hydrogen shell some of which is being burned into helium. Beyond this small core is the star proper, which becomes unstable and sheds its outer atmosphere to become a planetary nebula, while the remaining star contracts to become a white dwarf.

present state. As a red giant it will have a mainly carbon core surrounded by a helium-burning shell which, itself, is surrounded by a hydrogen-burning outer shell. At this point, the normally tight gravitational regulation of the star becomes much looser in the outer layers and begins to overshoot. When the outward-flowing radiation flux induces an expansion, it is no longer immediately suppressed by gravity but causes a significant expansion which is then followed by a contraction. These pulsations in the Sun will have a period of about 10000 years and they will grow in amplitude until the outer layers lift off from the star at a gentle velocity of about a few kilometres per second, leaving behind a hot core containing about 60 per cent of the Sun's original mass. Thermonuclear reactions will continue until its nuclear fuel is exhausted, when it will have been heated to a white-hot temperature of about 30 000° and be composed mainly of carbon surrounded by outer layers of helium and hydrogen. The Sun will then be a *white dwarf*, having contracted to the size of the Earth, and will be surrounded by its ejected atmosphere which will glow in the ultraviolet light of the nucleus; this shell will be a *planetary nebula*. It will reach this point about a 100 000 years after leaving its red giant state, and since its internal temperature is not high enough to burn carbon further towards iron, it will have no further nuclear energy resources, and will simply cool with time, becoming a yellow dwarf, then a red dwarf and, ultimately, a black dwarf after a further several thousand million years. The Sun will no longer shine; its life will be over. (See Colour Plate 14.)

In these last phases of the Sun's life, it will be in a degenerate state. This is a state of matter quite different from what we experience on Earth, or indeed, in any of the normal stars. It happens when the force of gravity becomes so large as to overcome the inter-atomic forces, imposing densities on the medium that are so high that, on Earth, a sugar lump or thimble full would weigh a tonne! At such densities, nuclei are forced closer together than the normal dimensions of atoms, which means that atoms, as we know them, cannot exist, electrons cannot have orbits linked to nuclei but are squeezed out into a kind of electron soup. According to classical physics, the star would then collapse because its kinetic pressure could not withstand gravity as it cools, but quantum mechanics shows that another kind of pressure, called degeneracy pressure, takes over and is a consequence of the Pauli exclusion principle and can be calculated from Heisenberg's uncertainty principle (both principles are explained in Chapter 7 in the section on quantum mechanics). The exclusion principle states that only one electron can occupy one quantum energy state at any time; hence, as the star cools, the lower energy states are filled and additional electrons have to have higher energies and therefore sit on top of the others, like bricks in a brick wall. This can be quantified with the uncertainty principle, which states that the uncertainty in position of an electron multiplied by the uncertainty in its momentum is equal to Planck's constant. (This principle also applies to the atomic nuclei, but these are so much heavier that their uncertainties are relatively negligible.) The uncertainty in position is given by the average separation of the electrons which comes directly from the density; this allows the uncertainty in momentum and hence the range of energy in each quantum state, in which only two electrons (with spins anti-parallel) can exist. The next two electrons in the neighbourhood have to have an energy which is higher by that range, and so on. Quantum mechanical calculations show that the pressure of a degenerate gas depends only on its density, just as the strength of a brick wall does not depend on the temperature of its bricks. They also show that a sufficient rise in temperature, and therefore of momentum, raises the product of uncertainties in position and momentum above Planck's constant, thereby breaking the degeneracy. This explains the hiccup in the Sun's evolution outlined above at the point of the helium flash. Degeneracy has another unusual property in that the more massive white dwarfs are smaller than lighter ones, the total volume being halved if the mass is doubled.

Degeneracy is caused by a gravitational field overcoming the interatomic forces and this poses an immediate question as to whether gravity can ever become high enough to overcome the quantum pressure of a degenerate gas. The answer is yes and it was provided many years ago by the Indian–American theoretical astrophysicist, Subrahmanyan Chandrasekhar, in a brilliant piece of work. He showed that when a star exceeded a certain mass, now known as the Chandrasekhar limit, the degeneracy pressure could no longer support it and it would collapse. He was able to calculate the limit entirely in terms of the gravitational and atomic constants, and he derived a value of 1.4 solar masses, above which the star will collapse. But a further question is immediately posed: what happens to it after collapse, does it enter some other kind of state, not yet known, or does it collapse completely? This is a question that will be re-addressed in Chapter 12 in connection with one of the great post-war discoveries in astronomy.

In the case of the Sun and similar stars, there is now considerable confidence that the general pattern of evolution is well understood. But this confidence should be treated with some caution because of what has become known as the solar neutrino problem. As has already been related, the Sun is currently burning hydrogen in its core, a process in which four hydrogen nuclei (protons) are fused into one helium nucleus, which yields 26 million electron volts of energy plus two neutrinos. The energy generated takes about ten million years to diffuse through the Sun and finally escape from its surface as electromagnetic radiation; so when you feel the warmth of sunshine, it is energy that was generated ten million years earlier; however, the very weakly interacting neutrinos escape easily at the velocity of light or close to it. The solar models are structured so as to explain the radiative energy output of the Sun, but they also predict the neutrino output, which is therefore, in principle, subject to test. However, since matter is virtually transparent to neutrinos, they are extremely difficult to detect; nevertheless, an experiment was set up in the late 1960s by the American astrophysicist, Raymond Davis Jr, with the specific aim of measuring the Sun's neutrino flux. A very large tank containing about 400 000 litres of cleaning fluid was placed in a very deep gold mine in South Dakota. Cleaning fluid was selected because it contains chlorine (atomic number 17), one of whose isotopes (atomic weight 37) can interact with a neutrino to produce argon (atomic number 18 and atomic weight 37) with the emission of an electron; since the argon isotope is radioactive, it is relatively easy to detect. A deep mine was selected so as to shield the detecting tank from all other particles, such as cosmic rays, which would not penetrate the 1.5 kilometres of Earth but which the neutrinos would pass through easily. But, of course, they also passed through the tank easily and the actual nuclear encounters were so rare that the rate recorded was only about one every two days, the lowest signal rate so far in observational astronomy. It therefore took a long time to build up sufficient statistics to obtain a meaningful estimate of the solar neutrino flux. At the time of writing, the observations have continued for about 25 years and give a neutrino flux nearly four times less than predicted by the best solar models.

This discrepancy casts some doubt on the accuracy of the theories of the Sun's energy production, its structure and the predictions of its evolution. The error may lie in the astrophysics, that is, in the determination of the distribution of density and temperature inside the Sun, or in the nuclear physics, that is, the rates and types of neutrino production in the Sun's core or the rates of neutrino–chlorine reactions in the detecting tank of cleaning fluid. Whether it is either of these or a combination of both is not known at the present time, but it is of some human interest to note that in the scientific debates on the solar neutrino problem the astrophysicists are generally convinced that the error lies in the nuclear physics, and the nuclear physicists are equally convinced that the error lies in the astrophysics. Although this reflects the degree of confidence of each group within its own area, the final answer will probably come from that group whose subject is the one in error – which could, of course, be both of them.

The general pattern of the Sun's evolution, described and discussed above, is followed by all stars less than a certain mass, which is about eight times the mass of the Sun. (Higher-mass stars will be treated shortly). However, there is one major quantitative difference in the evolution of the low-mass stars compared to the Sun and this is in their time-scales, which vary markedly according to their mass. At the higher mass of eight times that of the Sun, the rate of evolution will be about a thousand times faster and will therefore be complete in about a hundred million years; at the lower range in mass of about one-tenth that of the Sun, the rate of evolution will be about a thousand times slower, and the full lifetime of a star will take an immense 10 million million years to complete. This is a thousand times longer than the present age of the Universe, so all such stars are still in their infancy. The reason for this wide variation in the lifetimes of the stars is the fact that the thermonuclear generation of energy is a very strong function of temperature; hence, although a heavy star has a greater reservoir of nuclear fuel, it burns it at a very much faster rate because of its higher internal temperature, and therefore uses it up far more quickly.

There are some differences in the evolution of low-mass stars, related above, between those formed at the start of the galaxy and which exist mainly in the halo and globular clusters, and those that have formed since, such as the Sun, mainly in the galactic disk. The former, called Population II stars by astronomers, are composed almost entirely of the primordial elements of hydrogen and helium, and the latter, called Population I, also have about 2 per cent of the heavier elements produced by the nuclear transformations of hydrogen and helium in the earlier stars and released into the interstellar space, from which the later stars condensed. This causes differences in the opacity to radiation in the two types of stars and, for the same mass, this results in different energy loss rates; but the effect on evolution is not great and need not be elaborated any further here.

The lifetimes of high-mass stars, defined as being in the range from about eight times to about twenty times the mass of the Sun, are much shorter than that of the Sun and other low-mass stars, as has already been explained. Such stars that are still existing must have been formed relatively recently and are therefore clearly of type Population I, since stars as heavy as that which were formed in the initial stages of the galaxy will now have completed their total lifetimes and, indeed, will have contributed their share of carbon, nitrogen, oxygen and the other elements heavier than hydrogen and helium, into the interstellar medium. But the high-mass stars, compared with the low-mass stars, show not only the large quantitative difference of a much shorter life-span, but a major qualitative difference which occurs in the later stages of evolution. On formation, a high-mass star will start burning hydrogen in its core (but at a much greater rate than in low-mass stars) and will ultimately exhaust its hydrogen, at which point contraction will increase its temperature, hydrogen-burning will extend to the shell around it, and the core will be then heated to the temperature needed to initiate the burning of helium into carbon. It is now in its stage of expansion towards a red supergiant, but does not undergo the dramatic increase in luminosity of a solar-mass star, the increase in radiating area being roughly compensated by the drop in surface temperature. At the point when the helium core reaches its ignition temperature of 100 million degrees, it exceeds this temperature quite quickly and does not, like a solar type star, collapse sufficiently to become degenerate, and nuclear burning of helium into carbon proceeds without the hiccup of a helium flash. The star then proceeds towards its red supergiant phase and reaches a stage similar to low-mass stars, in which a carbon core is surrounded by successive helium- and hydrogen-burning shells.

It is at this point that the great qualitative difference between the evolution of low- and high-mass stars occurs. A higher-mass star has to be supported by a higher internal kinetic pressure, thereby causing the temperature of the core to be increased to the very high value, about a thousand million degrees, needed to initiate the thermonuclear burning of carbon to heavier elements. This proceeds with the immediate by-products of nitrogen, oxygen and neon being themselves burnt in turn until the core reaches the ultimate state of being composed of iron. It is then surrounded by successive shells of burning elements, which include, on the way upwards, silicon, neon, oxygen, carbon, helium and hydrogen. This onion ring model is a little oversimplified, since the successive shells will not be sharply separated and dynamic effects will cause some merging of the various layers. However, the critically important factor at this stage of the evolution of a massive star is that the iron core is inert: there is no nuclear energy remaining. The core contracts, causing a further increase in temperature to several thousand million degrees, at which point the particle energies are enough to induce a wide range of nuclear reactions, all of which are endothermic (i.e. they take energy out of the system), whereas all the processes we have considered so far have been exothermic and have provided the energy that has driven the star. Iron is broken up into lighter elements, reversing the processes that resulted in the evolution to the present situation, and can also be fused into heavier elements, causing an extraction of energy from the system and a consequent very rapid cooling: the pressure drops dramatically and the core collapses catastrophically on itself at a speed almost reaching a quarter that of light. The inner part of the core, containing material equal to about one or two suns and already squeezed to a size of that of the Earth, is compressed a million times further to reach unprecedented densities several times greater even of those within atomic nuclei. At this point, the electrons and protons are forced together to form neutrons whose degeneracy builds up an immense nuclear pressure which stops the collapse and initiates an energetic rebound. In the meantime the outer layers above the core, whose supporting pressure has disappeared, are in free fall at supersonic velocities and collide with the high-velocity rebounding core to induce a further massive increase in temperature by shock heating, which results in more nuclear synthesis of the elements in the interface region of the collision. The rebounding energy of the core, and the shock energy generated by the supersonic collision with the infalling layers, are sufficient to blow off most of the star at a considerable velocity of about 10000 kilometres a second. It is important to stress that the physical processes

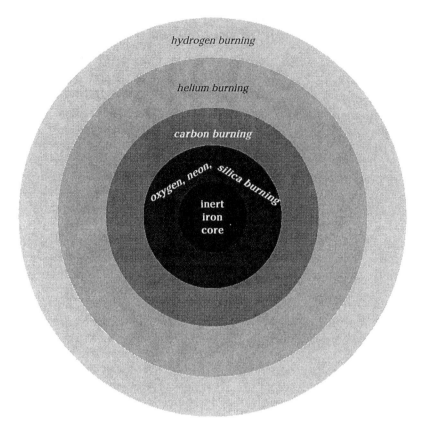

hydrogen burning

helium burning

carbon burning

oxygen, neon, silica burning

inert
iron
core

The core of a massive star (more than the mass of eight Suns) in a late stage of evolution when it is a red supergiant. The inner core has burned its hydrogen fuel into helium, then helium into carbon and all the subsequent heavier elements until it reaches iron which is inert. Beyond this are successive shells where some nuclear burning still continues in the oxygen, neon and silicon elements, then carbon, then helium and, finally, hydrogen; beyond that lies the bulk of the star which is not hot enough for thermonuclear processes to occur but still carries the outgoing energy flux. But since the inner iron core is inert and can no longer provide any nuclear energy, a catastrophic collapse occurs which results in the outer star being hurled into space in a spectacular explosion called a supernova.

outlined in this paragraph, which are completed in seconds, are far from being fully understood and are still the subject of intense astrophysical study today.

The explosion that occurs by the collapse of the inert iron core of a massive star in the final stage of its evolution is the cause of a supernova and is immense. Within seconds the energy output will equal a hundred times the total energy emitted by the Sun in its entire lifetime. The star destroys itself and, with the exception of a super-dense core, blows itself into interstellar space and forms an ever-expanding shell containing several solar masses of matter which contains, in addition to hydrogen and helium, large percentages of carbon, nitrogen, oxygen, and all the other naturally occurring elements from iron to uranium. Supernovae are the main source of the elements on Earth heavier than carbon, and for those heavier than iron, they are the only source. So, those many readers wearing gold or platinum rings should realize that those elements were synthesized entirely in the interior of massive stars which are now completely defunct.

Supernovae are rare events, since they occur only in massive stars which have relatively short lifetimes and therefore can only have been formed recently in astronomical terms. Only three have been recorded in our Milky Way galaxy over the past thousand years. The first, in 1054, was observed by Chinese astronomers and was so much brighter than Venus that it could be seen during the day; its remnant atmosphere is the Crab nebula and it has been an object of intensive study. The next two, as has been related earlier, happened in quick succession in 1572 and 1604, when they were observed as bright new stars by Tycho Brahe and Johannes Kepler respectively. There has not been a galactic supernova since, but the next best thing has happened: a supernova occurred in our nearest galaxy, the Large Magellanic Cloud (LMC), on 23 February 1987. (See Colour Plate 15.)

Despite the distance to the LMC of 170000 light years, the supernova was easily visible to the naked eye, the first time this had happened since 1604. The LMC is visible only from the Southern Hemisphere and the supernova was detected by astronomers in South America and New Zealand only hours, as it turned out, after its appearance. The announcement of the supernova initiated the most extensive set of astronomical observations ever concentrated on one object. It was subjected to study by the whole range of astronomical facilities available, both ground-based and space-based, and was to be observed in the visible, infrared, radio, ultraviolet, X-ray and gamma-ray regions of the spectrum; in addition, its neutrino flux was also detected, making a set of observations whose range is unique in astronomy for any one object. The result was a major advance in the understanding of supernovae, the final stage in the life of massive stars, in which many of the earlier concepts were confirmed and new problems emerged.

The optical brightness of the supernova continued to rise until it reached a peak on 20 May, 85 days after its appearance, when it was a bright object of third magnitude, and in the next two years it declined in intensity by a factor of a thousand. The increase in brightness was caused by the shock heating of the collapsing star by the energetically rebounding core, which resulted in the ejected shell which contains a mass equal to that of about 15 Suns. In that explosively shock-heated interface, the temperature rises to levels of several thousand million degrees, causing rapid nuclear synthesis, which burns all the lighter elements to iron (as in the core) and to even heavier elements as far as uranium. These are then ejected at high velocity by the shock, together with the other elements such as carbon, nitrogen and oxygen present in the outer layers. One of the abundant nuclei predicted to be formed in the shock region of explosive nuclear synthesis is that of the nickel isotope (atomic weight 58), which is radioactive and rapidly decays to the cobalt isotope of the same atomic weight; this is also radioactive and it emits gamma rays of known photon energy, thereby affording the possibility of an observational test of the theory. The gamma rays are heavily absorbed at the time of their formation, but as the envelope expands they will eventually diffuse outwards, with some degradation into X-rays by electron scattering, after about one year. Gamma rays can be observed only from space or at very high altitude but, fortunately, an American satellite with a gamma-ray detector was in orbit; this was the *Solar Maximum Mission* (SMM) satellite, which was pointed at the Sun, but supernova gamma rays could penetrate its envelope and excite the detector. Specially designed balloon payloads were also flown to observe the expected gamma rays. After about two years, enough data were accumulated to confirm the presence of cobalt 58 in the LMC supernova, a considerable boost in confidence of the understanding of the supernova event. Observations were also conducted in X-rays from the Soviet space station *Mir*, which carried appropriate instruments, and from the Japanese X-ray satellite *Ginga*, which happened to be launched just prior to the outburst. These were consistent with the expected degradation of the radioactive gamma rays from cobalt 58, but also showed other components that may be attributable to the heating effects of the expanding shell.

The programmes described above were the result of the concerted reaction of the international community to one of the most important astronomical events of this century. But there were two other sets of observations that were of crucial importance to the study of the LMC supernova and were unique to it. The first of these had already been made before the outburst; the spectrum of the progenitor star had been observed in the visible and ultraviolet, allowing a critical test of the theory of its evolution, which predicted that the inner core would collapse, triggering the supernova, when the star reached its red supergiant state. But the progenitor of the LMC supernova was a *blue* supergiant; it was hotter and smaller than predicted but with the same total luminosity. The star would pass through such a phase on the way to becoming a fully expanded red supergiant and, if the core did not collapse, it would return to such a state as it evolved. It would appear that the supernova was triggered either earlier or later than predicted by theory, depending on whether the full synthesis of iron in the core was completed before or after the star reached its red supergiant state. Initial considerations favour the latter and there are some observations in the visible and ultraviolet regions of the spectrum which support that view. These can best be explained by the existence of a circumstellar shell of about one light year in radius, which would have been ejected by the star in its red supergiant stage if the core had not collapsed, as happens for less massive stars in forming planetary nebulae. The existence of this circumstellar shell will be subjected to a definitive test when the rapidly expanding supernova outburst hits it at its velocity of about 30 000 kilometres per second, causing it to be visible directly in a somewhat spectacular explosion which should occur near the end of the century.

There are proposals to explain why the progenitor was a hot blue supergiant and not a cool red supergiant, but as yet none is fully convincing. The details of the final stages in the life of massive stars are therefore not completely understood, but the basic and central processes in a supernova explosion have now been adequately confirmed by the studies of the 1987 LMC event. This is particularly true of the second set of unique observations of the supernova, which were made in a serendipitous fashion by two experiments set up for a totally different non-astronomical reason.

In the post-war years, the discipline of particle physics made considerable progress in the study of the basic constituents of matter. This is not something to be described here except for one particular development which, in particle physics terms, turned out to lead nowhere. A theory had suggested that the proton was not permanently stable but would, in a sufficiently long time, spontaneously decay to more elementary particles. To test this, two experiments were set up, one in the USA and the other in Japan, to try and detect the proton decay over long periods of time. Since the proton is the nucleus of the hydrogen atom, the easiest way of detecting any decay was to monitor it in the form of pure water, which is cheap and transparent and whose molecules consist of two hydrogen atoms and one oxygen atom. Accordingly, the American and Japanese experiments consisted of huge tanks of pure water, surrounded by particle and photon detectors, which were buried in very deep mines in order to avoid competing effects such as cosmic rays. This they managed to do, but they could not shield their experiments from neutrinos and, although they detected no proton decay, they detected the neutrinos from the LMC supernova, the first non-solar neutrinos, the first truly cosmic neutrinos, ever detected.

Since the supernova neutrinos are emitted by the collapsed stellar core, which reaches temperatures of several thousand million degrees, they are very energetic, ranging up to more than ten million electron volts, and were detected via two processes in the water tanks. In the first, a neutrino collides almost head-on with an electron in one of the water molecules, and the path of the high-energy recoiling electron is easily detected. In the second, an anti-neutrino (yes, even the neutrino has its anti-particle) is absorbed by a proton in one of the water molecules, causing it to transform into a neutron and a high-energy positron, whose path is also easily detected. However, there is one major problem which has been touched on every time neutrinos enter this story: matter is virtually transparent to them and only a very tiny fraction of interactions occur. On 23 February 1987, the Earth was subjected to an extremely intense burst of neutrinos in which each human being received a dose of some 100000 million high-energy neutrinos, and yet only two huge tanks of water, surrounded by sensitive detectors, managed to detect some of them, and even then, only a few. The Japanese experiment measured 12 neutrino events over 12 seconds, and in the American experiment, it was 8 events over 6 seconds, a total of 20. Nevertheless, it was sufficient to allow a measurement of the total flux, which was in general agreement with the predicted value for supernovae. Nearly all the energy, about 99 per cent, was emitted as neutrinos, the other 1 per cent providing the kinetic energy that hurled most of the star outwards, together with the radiation flux by which it was observed. Because the ultra-high temperature and pressure in the collapsed core reduces matter to its most elementary states, the theory predicts that the neutrino output will consist of equal quantities of neutrinos and anti-neutrinos but, unfortunately, the water tank detectors could not afford a test of this. They were not designed for this particular purpose and did not have the means of unambiguously identifying an electron event, which would be triggered by a neutrino, and a positron event, which would be triggered by an anti-neutrino. The only possible way of discrimination would be in the distribution of the directions of the emergent particles, because the electrons, produced in recoil, will tend to have a forward direction compared to the incoming neutrino, whereas the positrons, produced in a nuclear reaction, will tend to have a broad distribution in their directions. However, the small number of events, a mere 20, was quite insufficient to allow any conclusion from the statistics, but the particle physicists have great confidence in the prediction since it is based on the parity of matter that they observe at the elementary particle level. This question of parity will be revisited in a cosmological framework in the final Chapter 13.

The detection of the 1987 supernova neutrinos, described above, confirmed the most basic aspect of the theory of such violent events: that they are the result of a catastrophic collapse in the thermonuclear processed iron core of a highly evolved massive star. This great contribution to astronomy by particle physics did result in some return to the contributor. An important question in particle physics is whether the neutrino has a rest mass, that is, whether, when it is stopped or absorbed, it becomes pure energy or whether there is still some residue particle of matter. If it has zero rest mass, like the photon, it will travel at the velocity of light, but if it has a finite rest-mass, it will travel at some lesser speed depending on its energy. Hence, if they have zero rest mass, the supernova neutrinos will have travel times to Earth identical to that of light, and slightly longer travel times if they have a finite rest mass, these longer travel times depending on their energies. Since there is a range in energy of the emitted neutrinos, determined by the equilibrium conditions in the collapsed supernova core, then there will be a range of travel times which will therefore cause a spread in arrival times and, in principle, allow an estimate of rest mass. The reason why this method has feasibility is the distance to the LMC of 170000 light years which means that even a minute difference in velocity will greatly magnify any time difference over such a long period. The observed spread in the detection of neutrino events is 12 seconds and the rest mass can be calculated from this if the spread is due entirely to a finite rest mass. But other factors may, and are likely to, be involved, such as a non-instantaneous release of the neutrinos, which means that we can only say that the spread lies between zero and 12 seconds. Hence, only an upper limit to the rest mass can be calculated, so its true value can lie anywhere between it and zero. The value obtained in energy terms (i.e. mc^2) is 16 electron volts, but a much tighter upper limit of 5 electron volts has been established in the laboratory. This is a very small value, as is seen by comparing with the electron rest mass of half a million electron volts. This result, of importance to particle physics, is also important to cosmology, as will be seen in the final chapter.

The above outline of the evolution of massive stars is a good description for those whose masses range from about 8 solar masses to about 20 or 30. In that range, the major difference is that the heavier stars evolve much more quickly to their final supernova state. In the initial phases of hydrogen-burning in their cores, these stars eject matter in the form of winds, which increase in intensity for the heavier stars. The winds are a result of the high temperatures, which cause intense radiation fluxes that drive off the outer atmosphere. But the energies involved are not significant compared to the radiation flux, and the basic model of a static and, gravitationally bound star is still valid, albeit approximately. However, if we climb to higher masses above about 20 to 30 times that of the Sun, the winds increase in intensity until they can no longer be ignored in considering evolution. The star can no longer be regarded as a static object that is gravitationally contained and balanced by the internal pressures of gas and radiation; it is a dynamic system whose declining mass and internal conditions cause changes in its structure which require fundamental adjustments to the theory of its evolution.

The main reason for these heavy stellar winds is the very intense radiation fields that are produced by the very high-energy rates of nuclear burning. The resulting radiation pressure is so great that it drives the material in the star's outer envelope into interstellar space; the photons are sufficiently numerous and energetic to propel the atoms beyond the pull of gravity. However, this radiation process is insufficient in itself to explain the strongest stellar winds observed, and some other additional phenomenon is also needed, perhaps involving some dynamic effect, such as a pulsation, which can lift the material to a point where the radiation pressure can take over.

The stars with the strongest winds are the Wolf–Rayet stars (named after their nineteenth-century discoverers), whose velocities range up to about 3000 kilometres per second and whose mass loss rates can be as high as one solar mass in 100 000 years. Since their ages are of the order of one or two million years, they will have lost a significant fraction of their initial mass, which is believed to be in the range of 30 to 50 times that of the Sun, and those observed are now down to about 10 solar masses. Because of their dynamic complexity, the evolution of these stars presents a challenging problem to stellar astronomers. They will still follow a path dictated by the development of nuclear synthesis, starting with the thermonuclear burning of hydrogen into helium in the core and, when complete, followed by the burning of hydrogen in a surrounding shell and of helium to carbon, nitrogen and oxygen in the core. As this proceeds, the outer layers, composed mainly of hydrogen, are constantly being shed. This scenario is supported by the observations that show that the Wolf–Rayet stars are deficient in hydrogen but have anomalously high abundances of carbon, nitrogen and oxygen. The outer parts of the star had been removed by the strong wind to reveal the thermonuclear processed products of the core. But major uncertainties still exist in understanding the evolution of the Wolf–Rayet stars, particularly concerning the final stages of their life. At present, it is not known whether they will shed sufficient mass so as to subside quietly into a white dwarf or to take the explosive route to a supernova – whether they will die with a whimper or a bang.

The stars that we observe in the sky all have different ages. Some were born at an early stage in the history of our Milky Way galaxy when star formation was very rapid; others, like the Sun, were formed 5000 million years later; and some, like the massive stars in Orion, were formed relatively recently, within the past few million years. Understanding the distribution and numbers of different types of star, from the hot to the cool and from the dwarfs to the supergiants, has to take account of this. It is like taking an instantaneous snapshot of the human race: there are two sexes, different races, languages and cultures, but also a complete spectrum of ages from the very young to the very old. An analogous situation also pertains for the stars, with the added complication (although nothing could seem to be more complicated than the human race) that stars have widely different lifetimes, the more massive stars evolving very much faster, even though they have more energy reserves, than the less massive stars.

This situation can be very much simplified by considering stellar clusters. It happens that stars do not condense singly in isolation but usually form in clusters or aggregates from some immense interstellar cloud; these appear in different forms, such as globular clusters, open clusters and stellar associations (see Colour Plates 16 and 17). But the important point in considering stellar evolution is that they were formed at the same time and lie at approximately the same distance; this commonality of age and distance is an extremely important factor. The relative brightness of the stars gives their relative luminosities which, together with their colours or spectra, allow their type to be determined; this in turn allows their evolutionary state to be identified, whether they are in the hydrogen-burning main phase, or in the red giant state or even later stages. Since stars evolve at different rates depending on their mass, they trace out the evolutionary development of the cluster. At the time of formation, all stars would have started burning hydrogen in their cores; this is their main phase, and how bright and hot they were would depend only on their mass. With the passage of time, the most massive stars in the cluster will evolve fastest and proceed through their red giant phase and on to their white dwarf or supernova stage. Others will only have reached their red giant phase, others will be on the way to it, while some, the lightest, will still be in their main phase of hydrogen core burning. This allows the type of those stars just completing their core burning to helium to be determined. From stellar models of evolution this, in turn, allows the age of the cluster to be estimated. This is an important parameter for the study of stellar evolution, but it also resulted in an estimate of great cosmological importance. The oldest stars in the galaxy are those in the globular clusters present in the halo, and the great ages estimated for these lie between 12000 and 14000 million years. Since the globulars were formed at the beginning of the galaxy, this is an estimate of its age and it establishes a lower limit to the age of the Universe, which therefore has to be 14000 million years at least. This result has an important bearing on the modern theories of cosmology to be discussed in Chapter 13.

In the study of the nature and evolution of stars, other measurements were made which are of great cosmological significance – of the cosmic abundance of the elements. These are derived from the analysis of stellar spectra which, in addition to the physical nature of the outer atmosphere such as its temperature and density, allows its chemical composition to be determined. This is not an easy task and it requires the setting up of a model atmosphere and solving the problem of the transfer of radiation through it on the last stage of its passage from the nuclear burning core. The first result obtained from a measurement of the strength of a specific spectral line is the number of atoms, or ions, producing that line, and that number is usually a small fraction of the total number of atoms and ions, which must be known in order to determine the total abundance of that particular element. In the simplest case of hydrogen, the most commonly observed spectral lines are those in the visible region of the spectrum and are caused by transitions from the second quantum level to higher states, yet most of the hydrogen atoms will have their electrons sitting in the first quantum level, which is the ground state; also, many and sometimes most, will be ionized and produce no spectral lines for observation. The problem therefore requires the total number of atoms in any one species to be determined from the numbers derived for a few individual levels, and this needs a theory of excitation, which is provided by quantum mechanics. This must be carried out for every ionization stage of the element, which is only two for hydrogen and three for helium, but which numbers more for heavier elements. Since some of these ionized species cannot be observed, an ionization theory is also needed. Given that all of these problems are solved by the application of quantum physics to model atmospheres, the chemical abundances of the elements can be determined. The reader should remember that these estimates refer only to the outer stellar atmospheres; as we have already seen, the inner parts of the star are undergoing nuclear transmutations. Determinations of chemical compositions are not confined to stars but are carried out in most astronomical systems, such as diffuse and planetary nebulae, but the models are very different and the physics has to be adjusted accordingly.

For the Sun, the abundances of the elements are now well determined and are referred to as "solar abundances". Broadly, these give the following values by mass – hydrogen 74 per cent, helium 24 per cent, carbon, and all the heavier elements 2 per cent. To get the abundances in terms of the relative number of atoms, the above values need to be divided by the atomic weights, which makes hydrogen even more dominant numerically. The solar values are very similar to those for the giant planets, which is not surprising since they were formed from the same pre-solar nebula as the Sun; the reader will remember that it was the Earth and the other terrestrial planets that condensed selectively out of the heavier elements so that the hydrogen and helium were severely depleted. When these studies were applied to the stars in the disk of the Milky Way galaxy (the so called Population I stars) it was found that their chemical abundances were the same, within the errors of estimation, as those of the Sun. There were some exceptions, such as the Wolf–Rayet stars, but these were understood, as we have seen, as objects in an advanced state of evolution in which the products of their nuclear burning have become exposed by the ejection of their outer atmospheres. Measurements of cosmic abundances in the interstellar medium, out of which the stars form by gravitational contraction, and in diffuse nebulae, which are regions of recent star formation glowing in the ultraviolet radiation of the young hot stars, also reveal the same abundances as in the Sun.

At this point, the reader might think that the cosmic abundance of elements is uniform throughout the Universe, with the exception of highly evolved stars, and that the solar abundances therefore indicated the composition of the primordial material out of which the galaxy was formed. But this is not the case; the commonality of solar abundances is true of stars, like the Sun, which were formed after the initial high rate of star births that marked the start of the galaxy itself. The most massive of these went rapidly through their life-spans, generating elements beyond hydrogen and helium by thermonuclear synthesis and hurling these into interstellar space to enrich the medium from which the Sun and other Population I stars were later to condense from.

The oldest stars in the galaxy lie in the halo above the plane of the Milky Way, mainly in globular clusters, and had long been recognized spectroscopically as being different from the Population I stars in the disk, and were designated as a different class, Population II. The main difference was the very weak presence or even absence of spectral lines caused by elements other than hydrogen. This led to the early conclusion in the immediate post-war years that the primordial material of the Universe was composed of pure hydrogen. However, the absence of helium was attributable to an effect that did not allow it to be observed; the helium atom is very tightly bound (see Ch. 7) and has to be subjected to high temperatures before it can be sufficiently excited to produce spectral lines in the visible region of the spectrum. But since the Population II stars are very old, near to the age of the galaxy, the massive hot blue stars have all completed their life histories and the remaining cool stars cannot excite helium to the point that it can be observed. This situation changed in the 1960s when a Population II planetary nebula was discovered in a globular cluster. The reader will remember that a planetary nebula is the blown-off atmosphere of a red giant star whose core has burned from hydrogen to helium to carbon; on its path to becoming a white dwarf, the remaining central nucleus becomes hot enough to excite the helium in the nebula and allow it to be observed. The subsequent analysis led to the conclusion that the original material from which the galaxy condensed was composed of 75 per cent hydrogen by mass and 25 per cent helium. Since it is believed that the Milky Way galaxy was formed (like all the other galaxies) in the early stages of the Universe, these values represent an estimate of the primordial abundance of the elements, based on observations, and of which all cosmological theories of the Universe have to take account, as will be seen in Chapter 13.

CHAPTER TWELVE

The Great Post-war
Astronomical Discoveries

There are more things in
Heaven and Earth,
Horatio,
than are dreamt of in your philosophy
Hamlet

The opening up of the whole electromagnetic spectrum to astronomical observations in the post-war years led to several exciting and unexpected discoveries that exceeded the hopes even of the pioneering scientists involved and took astronomy into a new era. It was not just a matter of saying "let us look and see"; the formidable costs involved precluded this as a justification but, with hindsight, it would still have paid off handsomely. When radio astronomy was initiated on the basis of the wartime development of radar, it was known from Karl Jansky's work in the 1930s that there were radio emissions from the galaxy. For the first of the space astronomies, ultraviolet, it was expected from optical observations that many celestial objects, particularly the hottest stars, would emit in that region, which is very important because it embraces the most important spectral lines of many of the astronomically common elements. It was also expected that significant emissions would occur in the infrared region, but the technical difficulties outlined in Chapter 9 delayed its exploitation for astronomy until 1983. In the early post-war years, it was not anticipated that there would be any significant emissions from celestial bodies in X-rays, and this region was only opened up after unexpected clues to its importance were afforded.

The first great post-war discovery occurred in the early 1960s when the very first quasar – a new and very different kind of galaxy – was located and identified. The radio telescopes at Cambridge had conducted surveys of the sky and this led to the third and most exhaustive catalogue in the late 1950s, which gave the positions and intensities of several hundred radio sources. The identification of these was a very important activity, but also a challenging one because of the limited positional accuracy of several minutes of arc, a field of view that would include several optical candidates for the radio source. Plausible identifications were then considered on the basis of whether any of the optical sources were different or peculiar, but this approach could be by-passed in the case of the 273rd source listed in the third Cambridge catalogue (3C 273), because it lay in the ecliptic plane and was therefore subjected to occasional occultations by the Moon, which blots it out as it passes in front of it. Such occultations have long been a tool in astronomy for determining precise positions because, by measuring the exact time that any source disappears as the Moon passes over it and the exact times at which it re-emerges, its extent and position can be determined, because the Moon's size and location are known precisely. Any ambiguity as to where it lies along the lunar limb can be totally removed by additional occultations when the Moon has moved against the celestial background to give a very precise measurement of the extent and location of the source. Observations were made of three occultations of the radio source 3C 273 with the Australian Parkes radio telescope, which showed that it was point-like and tied down its position precisely to a faintish blue star which had not received any previous observational attention other than to be recorded photographically in an optical sky survey conducted by the American Palomar observatory. A bright cosmic radio source had been unambiguously identified for the first time, but it did not seem to have any other distinguishing or peculiar characteristics. It was one of a host of similar blue stars and, as such, seemed quite unexceptional, but this innocuous and none too bright star, singled out only because of its radio emission, was to be shown to be one of the most exceptional and exciting objects in the Universe.

The developments so far had been entirely confined to radio astronomy; the source had been detected and recorded by the Cambridge group under Martin Ryle, Bernard Lovell's telescope at Jodrell Bank had shown that it was not an extensive source, and the Australian Parkes telescope under E. G. ("Taffy") Bowen had located it precisely. The next step was clearly to observe its optical spectrum, but the astronomy communities of the two countries involved thus far, Britain and Australia, did not have access to an optical telescope large enough for that purpose. The coordinates of 3C 273 were passed on to the Palomar observatory, and its spectrum was observed with the 200-inch telescope in 1960 by the Dutch-American astronomer Maarten Schmidt. This showed a pure continuous spectrum devoid of absorption lines but with strong emission lines superimposed on it. The lines could not be identified with any known element or species and since the state of spectroscopic information, which had been compiled extensively since the nineteenth century, was highly advanced, this posed a major and puzzling problem. The spectrum, recorded photographically, lay in an archive for three years when it was realized – in one of those flashes of insight that makes the discoverer ask "Why did I not think of that before?" – that the lines were very common in astronomy and very familiar to all astronomers. They were well known transitions in hydrogen and oxygen, as could be seen from their pattern, but their wavelengths did not lie in the normal positions because they were redshifted by a large amount. The redshift was the same for each line and therefore consistent with a Doppler velocity of recession. This was so high that it had confused an identification of the emission lines, which would otherwise have been trivial. The result astounded the world of astronomy; 3C 273 was travelling outwards at a speed of 44 000 kilometres per second, 16 per cent of the velocity of light, a much higher value than had been recorded in any other celestial body. Since such a velocity was inconceivable for an object within our own galaxy, it could only be the result of the expansion of the Universe and this enabled its distance to be determined. Immediately 3C 273 became the most distant known object in the Universe at about three thousand million light years, and also the most luminous, emitting more than a hundred times the flux of the Milky Way and other luminous spiral galaxies, even though it was much more compact. A new type of galaxy had been discovered of a very different nature to the spirals, ellipticals and irregulars that had been known of previously; it was to be called a *quasar* from a contraction of the descriptive term, *quasi-stellar object*. These developments, in which the final triumph (which fell to American astronomers) was denied to the British and Australian astronomers, who had located the object, because of their lack of a major optical telescope, was one of the factors leading to the building of the very successful Anglo–Australian telescope.

The identification of the nature of the first quasar, 3C 273, presented an immediate energy problem. How could such enormous energy be generated by a relatively compact body? Its size could not be determined because it appeared point-like in the sky, but this placed an upper limit on how large it could be and this meant that it had to be less than 1 per cent of the volume of the Milky Way galaxy (future developments were to show that it was very, very much smaller again). So, here was a body emitting more than a hundred times the energy of our own galaxy from a volume that was less than 1 per cent; what was the energy source? A clue was given by the compact nature of the source. It was clear that thermonuclear energy, the burning of hydrogen into helium, which had been so successful in explaining the immense energy of the stars, was inadequate and an even more powerful energy source was needed; in principle, only gravitation could provide this. Gravitation is a quite familiar source of energy; the rain that falls on mountains and into rivers has gravitational energy which can be tapped for human purposes in hydroelectric dams (but, of course, the original energy source is the Sun's radiation, which evaporates the oceans to provide the rain); gravitational energy is also released in the collapse of gas and dust in the interstellar medium and, if sufficient, will heat the condensing object to a temperature high enough to ignite thermonuclear reactions and thereby form a star. But the magnitude of gravitational fields then known to astronomers (of which the white dwarfs were the highest) was totally inadequate to provide the kind of energy demanded by quasars; the fields had to be high enough to exceed even the strong nuclear force (which provides the thermonuclear energy of the stars). This takes us into a new regime of physics, one in which gravitation can only be treated by general relativity. The discovery of the first quasar marked the birth of relativistic astronomy.

The energy processes that occur in the release of gravitational energy in the relativistic domain are very complex and are still not fully understood, but their broad characteristics are believed to be known. In the case of a quasar, it is proposed that the power source is a central black hole, that is, a concentration of matter sufficient to produce a gravitational field that prevents any particle escaping, even if it had the velocity of light (which is not possible by the special theory of relativity, except for a particle with zero rest mass, such as the photon) and therefore even traps light itself. The condition to have a black hole was worked out in 1916 by the brilliant German theoretician, Karl Schwarzschild, using Einstein's theory of general relativity (he was killed shortly afterwards on the Russian front during the First World War). The boundary of the black hole represents an event horizon where time stops and beyond which no communication is possible. The central black hole in a quasar is fuelled by material in its neighbourhood, in the form of stars, gas or dust, which are captured by the strong gravitational field and spiral inwards at ever increasing velocities, forming what is known as an accretion disk. Any matter that falls into the black hole cannot communicate and therefore cannot release any energy outwards. Infalling material can only release energy – by dissipative processes such as friction, viscosity and collisions – before entering the hole. Being a black hole, the velocities reached are a significant fraction of the speed of light and the energy generation can consequently be enormous. As much as 30 per cent of the rest mass energy (mc^2) can be generated compared to the 1 per cent available by the thermonuclear fusion of hydrogen to helium and ultimately to iron in stellar interiors. Apart from delivering the highest energy generation, the relativistic gravitational engine differs from all other energy generators in not needing a specific fuel; wood, coal or oil are needed for conventional burning or ignition, uranium is needed for nuclear fission reactors, and hydrogen will be needed for the future fusion reactors now under research and development. A relativistic gravitational generator, on the other hand, can be fuelled by matter of any kind. In principle, this makes it the most attractive source

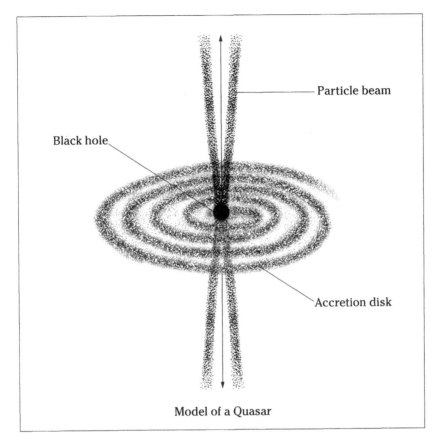

Particle beam

Black hole

Accretion disk

Model of a Quasar

A quasar is a galaxy whose energetics are dominated by a central black hole of mass about 1000 million Suns and an extent of several light hours or the dimension of the Solar System. The rest of the galaxy provides the black hole's fuel in the form of gas, dust or stars, which spiral inwards in an accretion disk at increasingly large velocities approaching that of light and generating energy due to friction and viscosity. Temperatures reach many millions of degrees, emitting X-rays as well as visible light with an immense energy output, which can be more than a hundred times those of the most luminous normal galaxies. The core radiation dominates, making the galaxy appear as a point source like a star. In addition to the radiation output, another source of energy loss is caused by jets of very high energy particles, which are emitted through the poles perpendicular to the plane of the accretion disk.

of energy for human use, since it would be inexhaustible and could be fed by rubbish and dangerous waste products. But the technology of constructing a black or near-black hole and controlling this as an energy generator is completely beyond human capability, now or into the future. Only on astronomical scales is it possible.

Since a quasar will have some angular rotation on formation, the infalling material will form an accretion disk with ever-increasing velocities as it approaches the centre, and will be heated to very high temperatures as the enormous energies are dissipated by friction and viscosity. This causes the very high radiative flux by which 3C 273 was observed. But another major form of energy loss is caused by the generation of relativistic particle beams which are emitted along the poles of the accretion disk. The particle velocities approach that of light itself, but the mechanism that accelerates them is not fully understood. A jet is present in 3C 273 itself, as was revealed by more detailed observations after its discovery. (See Colour Plate 18.)

Since the discovery of 3C 273, studies of quasars have been extensive and several thousand have now been detected and studied. The original model of their nature has been confirmed and made more detailed. The central black hole cores contain about a thousand million solar masses or about 1 per cent of the mass of our Milky Way galaxy. Typical sizes are several light hours, which is about the total extent of the Solar System out to the planet Pluto. Beyond the core, the outer regions of the infalling material, which is being gravitationally captured, is observed as it glows in the radiation emitted from its base near the black hole boundary where the energy generation is intense. This region extends over a few light months, still small compared to the separation of stars in our own galaxy, and shows typical velocities of about 10 000 kilometres per second. Beyond that region, gas is observed but at much lower velocities. No stars are observed but the presence of high velocity, relativistic jets emanating from the poles of quasars, has been fully confirmed.

Of the thousands of quasars now discovered, it has been established that they are most numerous at a distance and time when the Universe was half of its present age. The reason for this is not known, but then the whole question of galactic formation and evolution is one of the least understood areas of astronomy.

One of the astronomical games that developed in the search for new quasars was to find the most distant one. In this, the Anglo–Australian telescope has been particularly successful, greatly helped by its attendant wide-field telescope (the UK Schmidt), which was ideal for surveying the sky and identifying quasar candidates. The work may have been stimulated by the fact that it was British and Australian astronomers who missed the final triumph of discovering the first quasar because of the lack of a large telescope. At the time of writing, the most distant quasar, and therefore the most distant object in the Universe, is receding at a speed greater than 90 per cent of the velocity of light and is seen backwards in time to a distance where the Universe was less than 10 per cent of its present age.

Attempts to measure the distance of astronomical objects and the size of the Universe are, after more than 2000 years, nearing completion and it is appropriate to summarize them now. They started in the third century BC when Eratosthenes measured the size of the Earth, allowing Aristarchus to determine the distance to the Moon. These were good measurements, but subsequent estimates of astronomical distances were consistently much less than the true value; the Universe was immensely larger than could be envisaged. The size of the Solar System was finally established by Cassini and Richer in the seventeenth century, and the distance to the nearest stars had to wait until the nineteenth century, when independent measurements were made by Bessel, Henderson and Struve. Not until the first part of the twentieth century was a true scale of the Milky Way galaxy established by Shapley, and then Hubble determined the distance to the nearest spiral galaxy in Andromeda and went on to make the first estimate of the size and age of the Universe in the 1930s from its expansion. At that time Hubble's value was also an underestimate (by nearly a factor of ten) and subsequent improvements have caused it to increase until it has levelled out at the present time. However, the actual size and age of the Universe has still not yet been established with precision, because the actual rate of expansion (Hubble's constant) is still uncertain, and to calculate the age from it requires the setting up of a model Universe, which involves assumptions about its nature. The age of the Universe probably lies between about 10000 million and 20000 million years, and this matter will be addressed again in the next chapter on cosmology.

Since quasars are the most distant objects known, they provide the best means of probing the Universe and observing any material that might exist in between the galaxies in intergalactic space. In the study of quasar spectra, it was found that the more distant objects showed systems of discrete absorption lines which fell into two categories. The first was caused by strong and well known resonance transitions in common elements such as carbon and silicon, which were stripped of some of their outer electrons to become multiply ionized. These were identified unambiguously by the pattern of their wavelengths, which showed the identical Doppler shift and hence the same velocity. This was always less than that of the quasar, indicating the presence of a high-temperature cloud lying in the line of sight at an intermediate position between us and the quasar. Several such systems could be present in the spectrum of a single quasar, so many such clouds can lie in one line of sight at distances that could be easily calculated from their Doppler shifts and hence their velocities, using Hubble's law. There was no immediate explanation for the existence of extensive, high-temperature intergalactic clouds, but one came from ultraviolet observations of stars in the Large Magellanic Cloud made with the IUE observatory satellite. These revealed the same kind of absorption lines as are present in quasar spectra and they demonstrated that our

Milky Way galaxy has an extensive halo at a temperature of about 100000°. This surprising discovery has yet to be understood, but it did give an explanation for the high-temperature clouds lying in the line of sight to quasars. They were the hot halos of intervening galaxies and, as such, were not of a truly intergalactic nature.

The second category of absorption line systems in the spectra of distant quasars is a set of lines that always lie at wavelengths shorter than that of the resonance transition (called Lyman alpha) of hydrogen. These are caused by the same line of hydrogen in intervening clouds, whose Doppler shifts give distances in the same way as the lines produced in the halos of intervening galaxies. The number of lines in these absorption systems is so great that they have been called the *Lyman alpha forest* and they show that these clouds are far more numerous than intervening galactic halos. They are also quite different in that no absorption line of any other element can be associated with them; as far as the observations are concerned, they are composed purely of hydrogen. The clouds are truly intergalactic and are believed to be condensations in the early Universe which were neither large enough nor dense enough to form into a galaxy, and which must therefore be composed of primordial material. The observation of only hydrogen excludes all other elements, with the exception of helium, which is traditionally difficult to detect because its primary lines lie in the extreme ultraviolet. As will be discussed in the next chapter, primordial abundances are cosmologically important, and the search for helium in the intergalactic clouds is one of the prime aims of the Hubble space telescope.

Since the discovery of the first quasar, other types of active galaxy (i.e. galaxies with excessive energy generation powered by a black hole) have been realized to exist. Somewhat similar to quasars are the Seyfert galaxies, which were discovered in the 1940s and named after their discoverer (see Colour Plate 19). These have stars that show a spiral structure but are characterized by a very bright nucleus, and subsequent studies revealed some quasar-like properties such as very high internal velocities. It is now thought that the main difference between the quasars and the Seyferts is that the former have a far greater fuel supply, in the form of local material, to feed the black hole, and the central nucleus therefore dominates, making the rest of the galaxy relatively undetectable. Recent studies of normal spiral galaxies, which indicate a dense concentration of matter in their nuclei and some energetic effects, are leading astronomers to the belief that all spiral galaxies have a black hole at their centre and that the gradation from quasar to Seyfert to normal spiral galaxy is attributable to the availability of local fuel in the form of stars, gas or dust. Hence, spiral galaxies, like our own Milky Way, do not show the extremely active nuclei of quasars and Seyfert galaxies, because the fuel has been largely used up.

In some active galaxies, the major form of energy generation from the black hole is in the high-velocity jets streaming out of the poles of the infalling accretion disk. These are most easily detected at radio frequencies because the jets excite radio emission as they penetrate the surrounding medium; consequently, they are known as radio galaxies. Other active galaxies, which show evidence of a black hole by their intense energy generation and the presence of high velocity jets, are giant elliptical galaxies such as those in the constellations of Centaurus and Virgo.

Another major discovery was made in 1967 and came from an investigation being carried out by the radio astronomy group in Cambridge led by Tony Hewish. Hewish was interested in the nature of the interplanetary plasma caused by the solar wind as it streamed out of the Solar System, and he realized that this could be studied by observing point-like celestial radio sources, because these would appear to fluctuate in response to irregularities in the wind. A parallel optical effect is caused by the Earth's atmosphere, whose eddies and fluctuations make stars appear to twinkle, but not the non-point-like planets which appear quite steady. In order to conduct these studies of interplanetary scintillation, Hewish needed a large collecting area and built an array of dipole antennae covering more than four acres, optimized for detecting radio waves of about a metre in wavelength. The dipoles were aligned parallel to each other and could be tilted so as to observe sources at different declination as they passed through the field of view as a result of the rotation of the Earth. Only sources with angular sizes of less than two seconds of arc (like quasars) would display scintillation, and the first task was a survey of the sky to detect sources that were fluctuating. A key figure in conducting the survey was a young postgraduate student, Jocelyn Bell, who began to compile a sky map of scintillating sources. One of these appeared to be different from the others and unusual in its variability, leading to the belief that it might be a radio source whose fluctuations were intrinsic to itself. This caused Hewish to ask Bell to study the source further, but with the fastest time-resolution of the equipment, which had not been employed previously because the system had been optimized for the expected timescale of interplanetary fluctuations. This revealed the source to be a series of sharp pulses, with no detectable radiation between them, at a frequency of about one per second; the intensity of the pulses fluctuated in a seemingly random manner, but the time interval between them was exactly the same, like a high-precision clock. The first pulsar had been discovered.

If the object was a star, which was immediately and correctly supposed, it had to be of a uniquely different kind from all other known stars. The time duration of each pulse is only of the order of a millisecond, thereby placing a limit on the size of the emitting body. If, for example, the Sun pulsed instantaneously, the light signal from its central disk would reach us before that from its limbs, causing the signal to be of two seconds' duration (the Sun's radius is about two light seconds). In the case of the pulsars, the time duration of the pulses limited the size of the object to less than a few hundred kilometres. Further, a star that could pulsate or rotate in one second had to have an immense gravitational field to prevent it flying apart; it would have to be condensed to a diameter of about 10 kilometres. The white dwarfs, of which the Sun will ultimately become one, are stars that have contracted to the size of the Earth and whose internal conditions are consequently of a nature totally different from our normal experience, but for a pulsar we have a star that is so condensed that it could fit into the centre of any major city. Its internal nature would therefore have to be in a realm different from even that of the white dwarfs: it had to be a neutron star. This conclusion, reached quite quickly by the discoverers, was the result of earlier speculations by astronomers and physicists, who had considered the state of matter in a star that had used up its nuclear fuel but was heavy enough to collapse beyond the white dwarf stage. They concluded that the star would be very small and the material in it so dense that electrons and protons could not exist separately but would be squeezed into each other to form neutrons. It would be a neutron star, whose intense gravitational field was balanced by nuclear degeneracy pressure analogous to the atomic degeneracy pressure that supports a white dwarf and which was explained in Chapter 11. In this case it is the neutron and not the electron that has to find its own individual quantum state and others have to find higher quantum levels and therefore have higher energies, which build up like a wall to provide the pressure that supports the star. It is very difficult to convey the remarkably exotic nature of neutron stars and how totally different they are from anything we can remotely experience on Earth, but I will try with a simple numerical example: a sample of neutron star material the size of a sugar lump would have a weight on Earth of 100 million tonnes, equal to the total tonnage of *all* the world's shipping. Of course such a sugar lump of a neutron star could not exist on Earth or on its own because it would no longer be subject to the intense gravitational field that was compressing it.

Clearly, a pulsar is a neutron star and, since such an object had been considered theoretically, it might be argued that the discovery was not totally unexpected; but this is not so, because it was not believed that such small objects, devoid of nuclear fuel, would be observable. The fact that they are is attributable to the unique emission mechanism, which reveals them as sharp, bright pulses. Since pulsating or vibrating stars were already known (such as the cepheid variables) this was the first

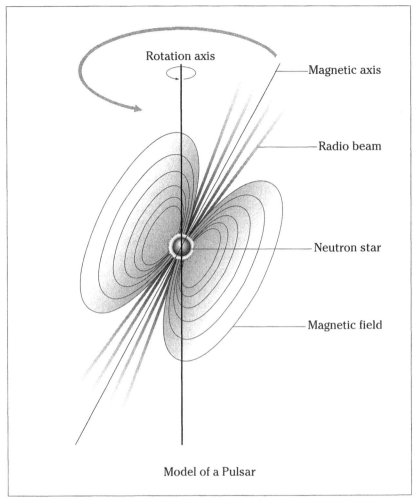

Model of a Pulsar

A pulsar is a star in a late stage of its evolution, which has collapsed to the point where gravity overcomes the nuclear forces of matter. Consequently, electrons and protons cannot exist freely but are forced into each other to form neutrons. It is a neutron star whose mass is greater than that of the Sun, but is so compact that its diameter is less than that of any major city! As a result of the collapse, its magnetic field and rotational velocity are increased immensely, but, as is usual in rotating bodies with magnetic fields (such as the Earth), the magnetic axis and rotation axis do not coincide, causing the former to precess about the latter at frequencies typically about 10–100 times a second. Since the main source of energy lost is by ultra-high energy particles, which can escape only through the magnetic poles where they generate beams of electromagnetic radiation, the effect is like a lighthouse, and, if the Earth lies within a beam, a rapid pulsed source of radiation is observable, which is the sign of a pulsar.

process considered, but was quickly abandoned in favour of a mechanism based on the rotation of the star. In whatever process leads to the star's collapse, angular momentum is conserved and causes a speed-up to the high frequencies observed. A similar effect will greatly intensify any magnetic field present in the body before collapse and, since neutron stars have conducting properties, these high rotations can produce intense magnetic fields of a dipole nature, like those of a bar magnet. Most rotating astronomical bodies, including the Earth, have dipole magnetic fields as a result of the rotation, but the axis of the magnetic field is tilted with respect to the rotational axis; in the case of the Earth, the tilt is about 11°. In neutron stars, any tilt in their magnetic axes will cause them to precess about the rotational axis at the rotational frequency. The next step in this argument, mainly due to Thomas Gold, is that a major form of energy loss is attributable to high-energy beams of particles (which are also exhibited by quasars), which escape from the *magnetic* poles only because, being ionized, they are inhibited from crossing the magnetic field lines and therefore travel along them. In so doing, they spiral about the magnetic field and emit synchrotron radiation (because they are decelerated) in the direction of their motion. This causes a directed beam of radiation which, because of the precession of the field, sweeps the sky like a lighthouse beam. If the Earth lies within the range of that beam, it detects the neutron star each time the beam sweeps past it. But, quite clearly, most neutron stars will not be detected as pulsars because their narrow beams will strike the Earth only occasionally.

In their initial survey, Hewish and Bell discovered three more pulsars, and the announcement of the discovery triggered surveys by several radio telescopes, which led to several more pulsars being found. Of these, one was of particular and immediate importance: a star in the centre of the Crab nebula which was located precisely by optical observations that revealed it was pulsing at visible wavelengths with exactly the same frequency as the radio pulses. Since the Crab nebula is the still-expanding shell of a past supernova, this clearly implied that the pulsar was the remnant nucleus of that explosion. It also added weight to the proposition that the pulsar emission was synchrotron radiation caused by a beam of extremely energetic particles interacting with the ultra-high magnetic field, since such radiation covers a wide range from radio to visible wavelengths and even to X-rays and gamma rays. It is now generally agreed that neutron stars are the result of the catastrophic collapse of the inner cores of massive stars in the final stage of evolution, which result in supernova explosions.

The discovery of the Crab nebula as a pulsar came at a very opportune moment, since it was then an object of intense astronomical study. A supernova of AD 1054, it lay at a distance of about 6000 light years (and therefore actually occurred 6000 years earlier) and has a diameter of about 12 light years as we now see it. Optical observations showed extensive areas of bluish light interspersed with pink filamentary structures. The latter were recognized as gaseous emissions, mainly by spectral lines of hydrogen. The former showed a purely continuous spectrum whose radiation was polarized (i.e. the electromagnetic oscillations tended to lie in one plane); this was the signature of synchrotron radiation. The Crab nebula was being illuminated by beams of high-energy electrons interacting with its intrinsic magnetic field. But the Crab supernova occurred more than nine centuries previously and any high-energy electrons produced in that explosive event would have decayed completely by now. So, the source of the energy that maintained the Crab nebula's synchrotron emission was a major astronomical puzzle. It was answered by the discovery of its central pulsar; the high-energy electrons were being supplied by the remnant rotating neutron star, whose beams continue to sweep the nebula and maintain the energy source. This conclusion is supported by the observation that the Crab pulsar, whose period is 33 milliseconds, is slowing down at a rate of about 3 milliseconds a century; such a measurement is an indication of the tremendous accuracy afforded by pulsars because of their very precise, rapid and sharp pulses. The observed rate of slowdown can be easily translated into a rate of energy loss of the rotating neutron star and this is equal to the energy emitted by the Crab nebula, which is more than 10 000 times that emitted by the Sun.

The Crab is the youngest known pulsar and, although there have been visible supernovae since it occurred (Tycho's in 1572, Kepler's in 1604, and the most recent in 1987 in the Large Magellanic Cloud), none of these are seen as pulsars, presumably because the beams from their remnant neutron stars do not strike the Earth. The Crab was also the brightest of these supernova, being visible during daylight. When discovered, it was the fastest pulsar, rotating at 30 times a second, and this was understandable since pulsars slow down with time because of the energy loss in their beams, and the Crab was the youngest. But as the survey of pulsars proceeded, some were discovered with frequencies much higher than the Crab, with periods almost as short as a millisecond. These so-called millisecond pulsars are remarkable objects, heavier than the Sun yet rotating at frequencies approaching a thousand times a second. It was quickly realized that these could not be single neutron stars but must be in a close binary system in which matter from the atmosphere of the normal companion star was streaming out and being accreted by and spinning up the compact neutron star companion. The existence of such interacting binary systems had already been discovered from quite different studies to be discussed in the following section.

At the time of writing, the total number of detected pulsars is well in excess of 600 and, from this, it is estimated that the total number of neutron stars in our Milky Way galaxy is about 100 000. It is a little difficult to explain this number if neutron stars are only formed in the catastrophic collapse leading to a supernova. It is therefore possible that a white dwarf in a close binary system may acquire additional mass by accretion from its neighbour finally to exceed the Chandrasekhar limit of 1.4 solar masses and collapse to a neutron star state.

Pulsars are still the subject of intense astronomical investigation and they provide a means of studying a super-dense state of matter not possible on Earth. They can be regarded as astronomical laboratories for the study of new areas of physics and, in that context, one object has become of particular importance. This was discovered in 1974 with the Arecibo radio telescope by two American astronomers, Russell Hulse and Joseph Taylor, who have studied it continuously ever since. The pulsar had a period of 60 milliseconds (about twice as long as the Crab), but it did not show the precise regularity that was typical of pulsars. Further study elicited the cause as being a Doppler effect caused by the pulsar being in a binary system and orbiting an unseen companion with a period of eight hours. Extensive observations were made over many years with the superb precision that is afforded by pulsars, and this allowed the parameters of the binary system to be determined accurately. The pulsar and its companion had

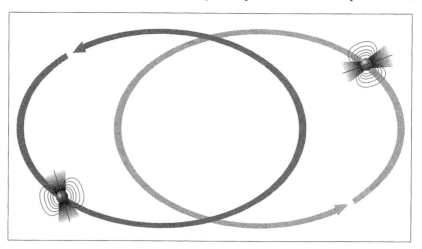

The binary pulsar consists of two neutron stars which are gravitationally bound to each other and have identically symmetric elliptical orbits as shown. The period of revolution is eight hours and each star moves through its orbit in a diametrically opposite position to the other. Although each star is 1.4 times the mass of the Sun, the scale of the system is only a solar diameter. Only one of the neutron stars, rotating at 17 times a second, is seen as a pulsar, since the beam of the other misses the Earth. The compact nature of the system, together with the remarkable precision of measurement that pulsars allow, have made the binary pulsar the best possible test laboratory for Einstein's theory of general relativity, which it has confirmed to an accuracy of one per cent.

the same mass and this was equal to the Chandrasekhar limit of 1.4 times that of the Sun. The separation of the two objects was very small, only 3 light seconds or less than a solar diameter. This meant that the unseen companion also had to be a very compact object and, with its mass, this could only be another neutron star whose pulsar beam did not sweep the Earth. It was a binary pulsar, a name by which it is still known.

The orbits of each star are almost identical but, of course, of exactly opposite phase and symmetry; they are also very elliptical, an important property for subsequent studies. The binary pulsar is a remarkable system, even by astronomical standards: two objects heavier than the Sun, contracted to the scale of a medium-size city to give the super-dense state of a neutron star and orbiting each other every eight hours, with separations that are only the diameter of a normal star such as the Sun. It is also a marvellous laboratory for conducting tests of Einstein's theory of general relativity; indeed, if the astronomical community had been invited to design a cosmic experiment for that purpose, it could hardly have bettered the binary pulsar that nature has provided. One of the early classic tests of general relativity was the advancement of the perihelion of Mercury (Ch. 7), but in the binary pulsar this effect is immensely greater because of the larger masses, the higher ellipticity and the greater compactness of the system. The perihelions of the two neutron stars advance at the rate of 4° a year, so they advance in one day what it takes Mercury a century to do. Measurements over several years, using the high precision provided by the companion (which is a pulsar), have given an agreement of 1 per cent with the value predicted by general relativity.

But a much more important result of the studies of the binary pulsar is a test, albeit indirect, of the one unconfirmed prediction of general relativity: the emission of gravitational waves. These are transverse waves, like light, in which the gravitational field oscillates at some frequency and, again like light, propagate at the velocity of light. The theory predicts that these are emitted by any body or system whose gravitational configuration changes and their intensity and frequency will depend on the degree and rapidity of the gravitational change. In the binary pulsar, which is a precise and orderly system, the rate of energy loss by gravitational waves can be calculated accurately. The only way this loss can be supplied is from the gravitational energy of the system; the two neutron stars would "fall" towards each other and their orbital period would decrease. Observations over 20 years, together with the superb precision of pulsars, allowed the detection of a minute rate of decrease in the period of 75 microseconds per year, which agrees to better than 1 per cent with the value predicted on the basis of gravitational waves from general relativity. The evidence for the existence of such waves is indirect, but the agreement of predicted and observed values is so close as to be very convincing. As the orbital period decreases, so will their ellipticities, until they become circular, when each star will be orbiting the other on diametrically opposite sides of a circle, and their gravitational wave emission will only be about 1 per cent of what it is now. Ultimately, the two stars will coalesce in another catastrophic explosion, but this will only happen after some thousands of millions of years.

At the time of writing, proposals to construct gravitational wave astronomical observatories are being actively considered by several countries. These would have the aim of directly detecting gravitational waves from emitting bodies and not inferring their existence from other effects, as in the case of the binary pulsar. The gravitational radiation from this object will be at frequencies determined by the orbital frequency of the binary system (once every eight hours) and its harmonics, but its intensity is well beyond any possibility of direct detection with present or foreseen technology. The detection of gravitational waves is a problem more difficult than even the detection of neutrinos and, unlike neutrinos, they can only be detected from an astronomical source. It is believed that the gravitational detectors being set up will only have the possibility of detecting the most violent astronomical events, such as a galactic supernova or the coalescence of two neutron stars. But when they come on line it may be that, once again, the totally unexpected discovery is made.

Another exciting and totally unexpected observation was made in 1962. Unlike the discovery of the quasars and pulsars, there was no immediate broad explanation, but it was to have major implications for the development of astronomy. Like the discovery of the first pulsar, it was the result of a project totally unrelated to what was discovered. At that time, the only astronomical medium known to be emitting X-rays was the Sun's corona, at a temperature of one or two million degrees. But this was a very rarefied medium and the intensity was very low, to the extent that, even assuming all solar-type stars had a similar corona, the possibility of detecting their X-ray emission was negligible. A proposal was then generated in the USA which was an attempt to detect the only other possible astronomical X-ray source – the Moon, which might be seen by X-rays, reflected from the solar corona in the same way that we see it in visible light by sunlight reflected from the Sun proper. The proposal was made by a group under the broad leadership of Bruno Rossi of the Massachusetts Institute of Technology (MIT), but included a key figure in Riccardo Giacconi. By space standards, the experiment was simple: to launch X-ray detectors above the absorbing atmosphere in the nose cone of an American Aerobee rocket, which, not being stabilized, would scan the sky as it toppled but, having quite wide fields of view, would probably encompass the Moon. The second launch in mid-1962 was successful and X-ray emission was detected, but, as subsequent pointing analysis showed, not from the Moon but from a broad region of the sky which peaked in the direction of the constellation of Scorpius. It was clear that non-solar cosmic X-ray radiation had been detected for the first time and its strength was such that it could not be related to stellar coronae but to some other, yet unknown, intense high-energy source. The experiment, whose aim of detecting solar coronal X-rays reflected from the Moon can now be regarded as somewhat mundane, made an unexpected discovery of great importance: it initiated the new discipline of X-ray astronomy, which has grown to be one of the most important areas of modern astronomy.

Shortly after the discovery of cosmic X-rays, another American experiment was launched using the same Aerobee rocket by Herbert Friedman's group at the Naval Research Laboratory. This had more sensitive detectors and better spatial resolution than the earlier flight and it revealed a very intense source in Scorpius, together with other weaker ones in the galactic plane of which one could be identified as the Crab nebula. The strong source was given the designation of Scorpius X-1 and it presented an immense problem of explanation. Its energy output in X-rays was ten thousand times greater than the total energy output from the Sun, the X-ray part of which is trivial. The typical temperature of the emitting body was 100 million degrees, so it could not be a normal star, and a neutron star, even at such a high temperature, would be too small to provide the high flux observed. Scorpius X-1 was to remain a major puzzle until X-ray astronomy moved into its satellite phase, eight years after its discovery.

The first cosmic X-ray source to be identified was the Crab nebula in 1964, before its discovery as a pulsar. Together with optical observations of the polarized blue emission (discussed earlier in this chapter), this led to the conclusion of the presence of high-energy electrons causing synchotron radiation in the local magnetic field. The later discovery of the Crab as a pulsar explained the existence of the high-energy electrons as being provided by its particle beams. This led the Friedman group to conduct another Aerobee rocket experiment, which showed that the Crab was an X-ray pulsar as well as an optical and radio one, with X-ray pulses 10000 times more powerful than the pulses in the radio region. Only young pulsars, like the Crab, pulsate over such a large wavelength range.

Towards the end of the 1960s, 30 cosmic X-ray sources had been detected, most lying in the Milky Way and therefore probably galactic in origin; but, apart from the Crab, these strong galactic sources could not be identified. Like the early radio observations, these X-ray sources could only be positioned with limited accuracy, so their identification with an optical image was uncertain, but their discovery triggered optical studies of the areas that contained an X-ray source, with its identification as the purpose. In the case of Scorpius X-1, improved X-ray data showed that the source had a small angular size and was possibly point-like, leading the optical astronomers to propose that it was a faint blue star in Scorpius. So, it appeared that the strongest X-ray source in the sky was an unremarkable star which had previously escaped any kind of detailed astronomical attention. The proposal was the right one, but its confirmation and explanation had to await the first X-ray satellite, which could provide the long observational times necessary but impossible with sounding rockets. This was an American satellite project conducted under the leadership of Giacconi and was launched from Kenya in order to achieve an equatorial orbit, thus enabling the spacecraft to avoid entering the high-energy particle belt that surrounds the Earth and is trapped in its magnetic field. It was launched on Kenya's Independence Day in December 1970 and in honour of that it was called *Uhuru*, after the Swahili word for freedom.

Uhuru scanned the sky and increased the number of detected X-ray sources to nearly 200. But it was also able to observe sources for extended periods of time, and an important and surprising result emerged when studying the third X-ray source in the constellation of Centaurus (Centauris X-3); it went out on a very regular period of just over two days. Another bright source in Hercules (X-1) was quickly discovered to also disappear on a regular basis, in this case in less than two days. This pattern was familiar to the optical astronomer as the signature of an eclipsing binary: two stars in orbit around each other, whose plane of motion lay in the line of sight to the Earth, causing each to be regularly eclipsed by the other and blocking out its emission. In these two systems, the X-ray source was disappearing as it passed behind its companion star. Optical studies had been made of an eclipsing binary star in Hercules, which had the identical period as observed in X-rays; identification was unambiguously established. A neutron star was orbiting about a normal star and this was fully confirmed by its detection as a pulsar with a period just over one second. The binary nature of the system was also revealed in the optical studies via the Doppler effect caused by the motion of the normal star, which is in an evolutionary stage similar to the Sun, but a little hotter, larger and heavier. The optical astronomers had known it was a binary but could not detect the unseen companion directly because it was emitting nearly all of its energy in X-rays; similarly, the X-ray astronomers could only see the X-ray emitting component of the binary system. Only when both sets of observations were available was the nature of the system understood and these were soon supplemented by ultraviolet observations in studies that clearly demonstrated the astronomical importance of observing over a wide range of the electromagnetic spectrum.

The identification of Hercules X-1 as a binary system involving a normal star and a neutron star presented an immediate problem: how are the X-rays generated from such a small body with such intensity at a level whose energy is about a thousand times greater than the total emission of the normal star? The clue lay in the very short orbital period of the system of less than 2 days, which can be compared with the 88 days that it takes the nearest planet, Mercury, to orbit the Sun. The two components are very close to each other, which results in tidal effects on the atmosphere of the normal star that cause material to rise until it escapes its gravitational field and falls into that of the compact companion. The system is a relativistic gravitational engine, in which the immense gravitational field of the neutron star is fed by fuel provided by its close companion. The material spirals down into an accretion disk at high velocities, which reach 10 per cent of the velocity of light, and, by friction and viscosity, heats up to tens of millions of degrees. Such temperatures are also reached in the interior of stars, where they induce thermonuclear reactions, but they are shielded by the body of the star and their energy leaks out very slowly (it takes about ten million years in the case of the Sun). In the X-ray binaries, as these systems have come to be called, the ultra-heated material is hardly shielded at all and can be seen more directly by its emission of X-rays. The very high gravitational fields result in an enormous energy output, which can be as much as 10000 times greater than the total energy flux emitted by the Sun.

Since these first developments, studies of interacting binaries have proceeded apace. These are systems with orbital periods of less than about 20 days and are therefore so close as to allow a transfer of material, by tidal and other effects, from one component to the other; the direction of flow depends on the nature of each star and its stage of evolution. Many close binaries have now been detected, but since they are signalled by being X-ray sources, they are systems in which one component has become highly evolved into a compact object and provides the gravitational engine fuelled by the material from its companion. Clearly, there are other close binary systems in which neither star has reached its final evolutionary state. The nature of the primary star varies considerably in the different binary systems; it can be a massive supergiant star whose stellar wind feeds the compact companion of its own volition. It can also be a normal star like the Sun, which transfers material from its atmosphere because of the massive tidal effect of the nearby companion. The compact companion in X-ray binaries is usually a neutron star, but in one case, Cygnus X-1, it is believed to be a black hole and is the strongest candidate of stellar mass for such a state; the other more well founded candidates are the massive black holes in the centre of quasars and other active galaxies. The primary star of the binary system Cygnus X-1 is a massive blue supergiant, and orbital analysis gives a mass for the secondary star of about ten solar masses, well above the expected limit of about three solar masses that can be supported by a neutron star.

Studies of interacting binary stars threw light on a related phenomenon that had puzzled astronomers for many years – novae – so called because they appear as new stars. A nova is a phenomenon totally different from a supernova and this was realized well before either object was even remotely understood. Although the increase in brightness of a nova can exceed factors of 10000, this is still minute compared to the immense output of a supernova; also, a supernova is a single event which then decays with time, whereas a nova, which also decays with time, can reappear. Each nova is named as such and designated by the constellation it appears in and the year of its outburst. One of the brightest recent novae was Nova Cygni 1978, which was studied extensively, particularly by the *International Ultraviolet Explorer* (IUE) satellite, and led to a broad understanding of the nature of novae. They are close interacting binary systems in which one component is a white dwarf and the other is a normal star. A transfer of material occurs from the atmosphere of the normal star to the white dwarf, and this orbits about it in an accretion disk, as in the case of the X-ray binaries. The amount of material, mainly hydrogen, in the accretion disk builds up until it reaches a stage where it becomes unstable and collapses in the gravitational field of the white dwarf, to strike its surface with a consequent release of gravitational energy. This is not on the same scale as in the X-ray binaries, where the compact component is a neutron star, but it is enough to heat the material to the many millions of degrees needed to initiate a thermonuclear runaway, which burns the hydrogen into heavier elements. It is like a hydrogen bomb in which the trigger is the gravitational infall and not the fission bomb of uranium or plutonium that triggers the man-made hydrogen bomb. The thermonuclear explosion is the cause of the nova outburst and its by-products were detected directly by IUE.

The evolution of close interacting binary stars presented a new and complex problem to the stellar astronomer. Whereas the evolution of single stars was becoming broadly understood, as, by inference, were stars in widely spaced non-interacting binary systems, the exchange of material in close binaries means that the masses of the stars are changing with time and therefore their rates of evolution are also changing with time. The heavier star will evolve more quickly, but in an exchange of matter its companion can start catching up. In the X-ray binaries, the neutron star can have resulted from a supernova explosion of a massive star, but another possibility is that it was a less massive star and that it became a white dwarf which accreted material from its companion to the point that it exceeded the Chandrasekhar mass limit and collapsed to a neutron star.

The evolution of close binary systems is an exciting and important area of modern astronomical study, and questions such as "How did the binary pulsar, two neutron stars in a very close orbit, evolve?" are of compelling interest.

Since the launch of the first X-ray astronomy satellite, others have been launched by the USA, the UK, Japan and Germany. The current position is that about 60000 celestial X-ray sources have been detected and that most of these have been identified. The strongest galactic sources are the X-ray binaries and pulsars, already discussed, but there are also many extragalactic sources which include the quasars and other active galaxies. It is clear that the X-ray sky reveals those objects with intense gravitational fields which are being fed by infalling matter to form relativistic gravity engines, black holes, neutron stars and, on a lesser scale, white dwarfs. Other X-ray sources, of much less intensity, have been detected from stellar coronae, like the Sun, which at the start of this subject were considered to be the only possible celestial sources of X-rays.

Gamma-ray astronomy was also opened up in the 1960s. Gamma rays are even more energetic than X-rays and have photon energies greater than about 100000 electron volts (100keV). The separation between gamma rays and X-rays developed historically, with X-rays being produced artificially in high-voltage discharge tubes, and gamma rays being observed from nuclear reactions in radioactive materials. However, the initiation of gamma-ray astronomy was not the surprise that marked the opening of X-ray astronomy, because it was anticipated that there would be gamma-ray emission from the plane of the galaxy. This was because of the existence of cosmic rays, exceptionally energetic particles that bombard the Earth's atmosphere and whose existence had been known since 1912. Since then, extensive studies have been conducted which showed that cosmic rays were composed mainly of protons and electrons, together with some nuclei of heavier elements, whose energies reached enormous values about a million times greater than the highest achieved in the giant man-made particle-accelerating machines. The source of the cosmic rays of lower energy is thought to lie in supernova remnants, but the nature of the highest energy particles and their method of acceleration is still a mystery. The energy of the cosmic ray particles is such that they are able to smash the atoms in our atmosphere completely to produce showers of subatomic particles and gamma rays. It was realized that similar effects would occur in collisions of the cosmic rays with the interstellar gas, and the consequent gamma rays could be observed from a satellite. The first detection was made in 1968 from a secondary package installed on the

American Orbiting Solar Observatory (OSO-3). This observed gamma rays from the plane of the Milky Way (where the interstellar gas is concentrated), with an increased intensity in the region of the galactic centre.

Another American space mission, the Small Astronomical Satellite (SAS-2) also carried a gamma-ray detector which confirmed the emission from the galactic plane but also detected two discrete sources in the direction of the Crab and Vela pulsars. The positional accuracy of the gamma-ray sources was poor (as was the case for the first radio and X-ray sources), but unambiguous identification with the Crab and Vela was possible because of the pulsed nature of the emission, which had the same period as the radio pulsations. The first satellite with a prime gamma-ray mission was launched in 1975 by the European Space Research Organisation and was coded COS-B. This produced the most accurate measurement of the gamma-ray emission from the galactic plane and confirmed the emission from the Crab and Vela pulsars. More importantly, it detected 23 additional discrete gamma-ray sources, of which only one could be identified as being the quasar 3C 273, whose gamma-ray power exceeds that which it emits in the X-ray or optical regions. The other 22 lie in the Milky Way and can be assumed to be galactic, but have defied identification with any other source observed at other wavelengths. The detection of discrete sources, completely unexpected at the initiation of gamma-ray astronomy, took the subject into a new era.

One of the most exciting developments in gamma-ray astronomy was the detection of gamma-ray bursters: intense flashes of gamma rays from discrete cosmic sources. These were first detected in 1967 *before* the observation of gamma rays from the Milky Way by OSO-3 and so, strictly speaking, can claim to be the first detection of cosmic gamma rays. However, the observations were made by US Air Force satellites searching for clandestine H-bomb detonations in space, and were therefore classified. Not until it was established that the events were cosmic rather than man-made were the data published in 1973. Since then, many observations have been conducted by satellites, of which the American Compton Gamma Ray Observatory is the most notable. Today, several thousand gamma-ray bursts have been detected; these last for up to several seconds and the strongest ones, at peak, cause a greater energy flux at the Earth than does any other object beyond the Solar System.

The nature of these intriguing phenomena is completely unknown and there are very few clues to guide us towards an understanding. No bursters have been identified with any astronomical object observed in other regions of the spectrum, but the error in position of a few degrees makes such identifications extremely difficult. The distribution of events is more or less uniform over the sky, implying that they are not galactic, since they do not lie preferentially in the galactic plane or the galactic halo. The gamma-ray bursters are, therefore, either extragalactic or they lie in a very extended halo of the galaxy; no data are available to distinguish between these two possibilities. If they are extragalactic, a major energy problem is posed as to what possible process could produce the immense gamma-ray powers that result. If they lie in some extensive galactic halo, it is conceivable that the lower powers required could be the result of phenomena involving neutron stars, such as the catastrophic merger of two such objects in a collapsing binary. But the problem of explaining such an extended galactic distribution of neutron stars is immense.

It is appropriate to end this section on gamma-ray astronomy with a brief summary. Emission from the galactic plane can be explained by the collision of cosmic rays with the interstellar gas, but the mechanism that produces the high-energy cosmic rays is completely unknown. The discrete gamma-ray sources that have been identified are either pulsars or active galactic nuclei and their production is clearly connected with a relativistic gravitational engine but the actual mechanism is not understood (in quasars, the power in gamma rays exceeds that in any other spectral range). The most puzzling phenomena of all are the gamma-ray bursters, whose explanation will probably have to wait for some major observational breakthrough. So, at the time of writing, this whole area of ultra-high-energy astronomy looks like a total mystery.

One of the most important techniques in astronomy is the method for determining the masses of celestial objects. In any gravitationally bound system that neither disperses nor collapses, the velocities of individual bodies must be such as to resist the force of gravity, and the mass can be determined from Newton's theory of gravitation by measuring those velocities and the size of the system. In binary stars, the Doppler effect gives the velocity and its oscillation gives the orbital period and hence the size of the system, enabling the masses of stars of different types to be established. A programme to apply an adaptation of the same technique to measuring the masses of galaxies and clusters of galaxies was initiated in the 1970s by Vera Rubin and her colleagues at the Carnegie Institution in Washington. Spiral galaxies such as our own are rotating systems, with the orbital velocities of stars around the centre preventing their infall. Doppler velocities can be measured from the centre out to the visible edge of a galaxy and even beyond by using radio observations of the 21 centimetre line of neutral hydrogen. The measurement at each radial point allows the mass within that radius to be calculated and these *rotation curves* therefore enable the mass distribution to be determined as well as the total mass of the galaxy.

251

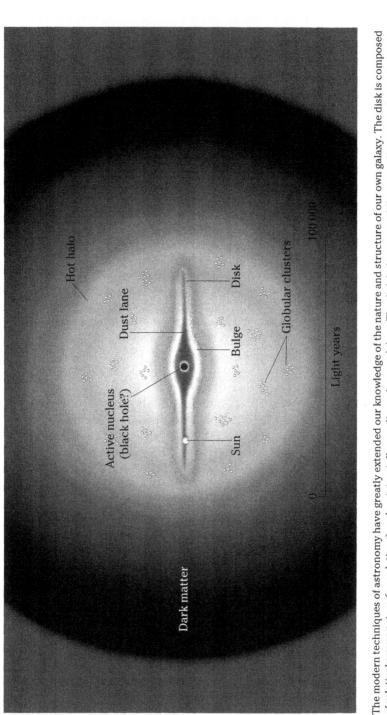

The modern techniques of astronomy have greatly extended our knowledge of the nature and structure of our own galaxy. The disk is composed of relatively young stars (population I) and an interstellar medium of gas and dust. The disk bulges around the central active nucleus which may contain a black hole. Around the disk is a halo which contains the older (population II) stars in globular clusters which are embedded in a hot gaseous medium at a temperature of 100 000° Kelvin, whose heating source is not known. Beyond that halo there is an even more extensive corona composed of dark matter whose nature is completely unknown and whose presence is revealed only by its gravitational effect.

The expected result was not forthcoming and nature yielded another major surprise for the astronomer. The masses derived were about ten times greater than the masses of all the stars and interstellar material that could be detected. Observations of clusters of galaxies, where the velocities measured were those of the individual galaxies, confirmed this result, which had also been inferred as long ago as 1933 by the Swiss-American astronomer, Fritz Zwicky. We could only see about 10 per cent of the Universe; 90 per cent was in the form of some unknown invisible material whose presence was detected only by its gravitational effect! The determination of the nature of this so-called dark matter is one of the most important challenges in astronomy. It is also critical to understanding the nature and evolution of the Universe and will be returned to in Chapter 13 on cosmology.

The distribution of dark matter in spiral galaxies showed that it extends well beyond the visible disk, that it is well beyond the stars, and that it is roughly spherical in its distribution. Together with the observations in the optical, ultraviolet, infrared, radio, X-ray and gamma-ray regions, this enables us to formulate an overall picture of our own Milky Way galaxy. It has a disk with a thickness of about 5000 light years, which contains the spiral arms and has a diameter of about 100000 light years. In the centre, there is a bulge which shows activity that may be the result of a central black hole, but one without a substantial source of local material as fuel. Around this is a spherical halo of stars containing the globular clusters with a radius about the same as the disk. There is also a hot halo of gas at about 100000° Kelvin of about the same extent as the globular clusters, which was discovered by ultraviolet observations of the Large Magellanic Cloud made with the IUE satellite. Beyond that there is the very extensive corona of dark matter reaching out to about 200000 light years.

The last major discovery to be reported in this chapter was made in 1979 by Dennis Walsh and his collaborators and was the result of a long-term programme at Jodrell Bank in Cheshire to identify radio sources. One of these was found to be coincident with two faint blue stars separated by only six seconds of arc. Optical observations revealed that they were two quasars with identical redshifts (the shift in wavelength divided by the wavelength) of 1.41 (corresponding to velocities of recession of 71 per cent of the velocity of light) and identical spectra; the only difference was that one was a little brighter than the other. The announcement stimulated observations with many facilities over a wide spectral range in the ultraviolet, visible, infrared and radio regions. These quickly established that the object was not a pair of twin quasars of the same age, same distance and nearly identical natures, but the double image of a single quasar formed by an intervening galaxy acting as a gravitational lens. The light beam was being bent around one side of the intervening galaxy by its gravitational field and around the other side, each producing an image slightly displaced from the other. The effect had first been demonstrated, on a much reduced scale, by the bending of starlight near the solar limb observed during the solar eclipse of 1919 as a test of Einstein's general theory of relativity (discussed in Ch. 7). The essential observational factor that came down in favour of the gravitational lens explanation of the double quasar was that the intensity ratio of the two images was exactly the same in the ultraviolet, visible, infrared and radio regions of the spectrum. The lens was a perfect achromat, an essential property of gravitational light-bending which comes from the basis of general relativity, the equivalence principle. This basis was established by Galileo 400 years earlier when he demonstrated that all objects, whatever their mass or composition, fell identically in a gravitational field. Hence, the heavy ultraviolet photons, with their high frequency giving them greater energy and therefore greater mass, will move in a gravitational field identically with the much lighter radio photons, with their low frequencies, and therefore low masses.

The discovery of the double quasar as the first gravitational lens phenomenon was very exciting but, unlike the first quasars and pulsars, it was not a totally unexpected event. The bending of light by a gravitational field had been known for some time and could be treated quantitatively with Einstein's general theory of relativity. With the discovery of the distant quasars, it was soon realized that any intervening galaxy would cause lensing effects such as the simple binary image of the double quasar, and a statistical assessment concluded that such effects should be sufficient in number as to be observable. Theoretical studies showed that the form of the imaging depended on the position and shape of the intervening galaxy. In the case of the double quasar, the intervening galaxy turned out to be flattened and slightly off-centre; this causes a double image, with one image slightly brighter than the other because of a greater magnification in its flight path. In the case of a galaxy that is precisely in line and spherical in shape, the light will be bent into a circle surrounding the galaxy and is called an Einstein ring.

Deep sky photographs taken with the Canada–France–Hawaii Telescope on Mauna Kea, Hawaii, revealed the intervening galaxy to be a giant elliptical with a redshift of 0.39 (corresponding to a velocity of recession of 32 per cent of that of light). This galaxy does not lie at the median point between the two images, but is closer to one of them. This means that the light from one image has a shorter path and therefore takes less time to reach us than the light from the other image. The time delay between the light from the two images is directly related to the Hubble time and the rate of expansion of the Universe, so if it could be measured it would offer the great prospect of determining the distance of the quasar from relatively simple geometry and thereby determining Hubble's constant (velocity divided by distance) to a distance far greater than any other method. Determination of such a time delay requires some identifiable event, or events, to occur in the quasar which can be observed in one image and then, after the time delay, in the other image. Such an event was observed in the ultraviolet with the IUE satellite, which saw a decrease in the brightness of the nearest image by a factor of two in September 1980, followed by a similar decrease in the other image 22 months later. These reductions in the ultraviolet emission, caused by variability in the power generation of the central black hole, are identified as the same event seen at different times in the two images, and gives the time delay of the system as 1.8 years. According to the model of Universe adopted, this will yield a value of Hubble's constant and the rate of expansion of the Universe, which is probably the most accurate estimate yet made and certainly the one to the greatest distance – about 10 000 million light years, when the Universe was about one quarter of its present age. The cosmological significance of this will be discussed in Chapter 13.

The major impact on astronomy caused by the extension of the observable spectrum from the narrow visible region to all other spectral regions is clear from the preceding sections of this chapter. To the relatively sedate Universe of the pre-war era consisting mainly of stars, has been added a range of violent objects exhibiting very energetic processes in quasars, pulsars and X-ray binaries; also, we have the great mystery of dark matter and the gamma-ray bursters. It is a very different and very exciting Universe, but one that presents major problems of understanding which are being tackled actively. But not all of the great post-war discoveries in astronomy have been presented above; one – the unexpected and almost accidental discovery of the cosmic microwave background – is of immense cosmological significance, and is discussed in the next and final chapter, on cosmology.

CHAPTER THIRTEEN

The Nature, Origin and Evolution of the Universe

To see a world in a grain of sand
And heaven in a wild flower,
Hold infinity in the palm of your hand
And eternity in an hour
Blake

The picture of the Universe that the astronomer had in the immediate post-war years was based on the observations made in the 1930s, particularly by Hubble, using the then largest telescope in the world, the 100-inch on Mount Wilson near Los Angeles. The main building blocks were stars that formed into massive aggregates called galaxies and these varied considerably in size and nature. Most were spiral in shape and contained, typically, about 100000 million stars. Others were elliptical in nature and others irregular (see Ch. 8). The galaxies themselves grouped into clusters containing up to a few thousand members; our own galaxy is a member of a small irregular cluster, called the local group, and the nearest large cluster lies in the direction of the constellation Virgo, and is therefore called the Virgo cluster. There is even a larger-scale structure in the Universe with clusters being linked in so-called superclusters, which cause depopulated regions called voids. But on the very large scale, the Universe seems to be equally populated everywhere and to look the same statistically in any direction. This had led to the adoption of the *cosmological principle*, which states that on the very large scale, the Universe is homogeneous, that is, it is the same on average at every position, and that it is also isotropic, that is, it looks the same on average in every direction.

Another fundamental property of the Universe, established by Hubble, is that it is expanding, that is, all galaxies are receding from each other at a rate proportional to their separation. An analogy to this is to take points on the surface of a balloon that is being inflated; two points separated by twice the distance of any two other points will appear to be separating at twice the speed. If the Universe is infinite in size, as is now generally assumed, any one observer will only see a finite Universe, out to a distance where the velocity of recession reaches that of light, when the light no longer has any energy for detection or, if the Universe has a finite age, out to the distance that light has travelled in that time. In either case an observer would seem to be at the centre of the Universe, but since this would be true of any observer in any part of an infinite Universe, the Copernican philosophy is not violated. Also, since the observable Universe is finite, Olbers' paradox (Ch. 6) is explained.

An important factor to bear in mind when considering observations of cosmological significance is that any object or phenomenon is seen as it was when it emitted the radiation we detect. We can observe the Universe only in a flash of space–time; we cannot observe it as it is *now*, but only at times in the past determined by distance. This is true of all observations, including those made on Earth, but here the distances are so small that light travel-times are extremely short and we see everything virtually instantaneously and do not have to think in space–time. But the scale of the Universe is so immense that the travel-times of light are very significant cosmologically; for example, the most distant quasars (see Ch. 12) are seen in the remote past when the Universe was only a tenth of its present age.

The cosmological principle and the expansion of the Universe formed the only basis for considering its evolution as a whole in the immediate post-war years. This was a remarkably limited foundation for tackling one of the greatest intellectual challenges to the human race, but Einstein's theory of general relativity allowed a consideration of the dynamical evolution of the Universe and showed that the observed expansion must be being slowed by the gravitational effect of the material within the Universe. Hence, in the past, the Universe had to be much more closely packed and to be expanding more rapidly. Extrapolating backwards in time takes us inevitably to a starting point, or a beginning, when the Universe would be infinitely dense and would then embark on its evolutionary course to the status we see today. This led to the big bang theory of cosmology, in which the Universe originated in an immense explosion at a *finite* time in the past. It was introduced in 1948 by George Gamow, who argued that it would be immensely hot as well as immensely dense and therefore be a seat of nuclear synthesis that might explain the cosmic abundance of all the elements. He was not right in this, since, as we have seen in Chapter 11, all the naturally occurring elements from carbon to uranium were processed in the interiors of stars, but his proposal became an important issue when the early primeval abundances of the elements that emerged from the big bang were considered.

A major problem with the big bang concept, philosophical as well as scientific, is that it requires an instantaneous creation, a point in time when the Universe, as an ultra-dense ultra-hot medium, came into being and before which there was nothing, no matter, no energy and no time. This was one of the factors that led Hermann Bondi, Thomas Gold and Fred Hoyle to propose an alternative theory in 1948, the *steady state* theory of cosmology. This argued from a basis of the laws of thermodynamics (Ch. 6), the first of which states that energy is conserved, that is, energy can change from one form to another, but the total amount is constant; the second law states that, in all energy transformations, the total entropy increases until it reaches a maximum or plateau. Entropy is most easily regarded as an indication of the amount of non-usable energy that is available; for example when water is boiled in a kettle and then used for whatever purpose, it ultimately cools down to the temperature of its surroundings, when the energy is still conserved but it cannot be used again, say to reheat the kettle. Hence, energy will continue to transform into unusable energy, and the second law of thermodynamics inevitably means that the Universe will run down and become dead. But stars are still shining and energy is still being generated and transformed; hence the Universe must have a finite age, otherwise it would have run down by now. There must therefore have been a beginning or a creation. This is met by the big bang theory in the form of a single instantaneous explosive event. But if a creation could occur instantaneously, then the steady state cosmologists argued that it was just as acceptable, intellectually, to have a continuous creation of matter, thereby eliminating the need to have a beginning of time.

The steady state theorists therefore proposed the *perfect cosmological principle*, which states that, on the large-scale average, the Universe is not only homogeneous and isotropic in space, but also unchanging in time; both the geography and history of the Universe are unchanging on average. But having reached this point from the basis of the second law of thermodynamics, it now became necessary to violate the first law by invoking the continuous creation of matter (and therefore of energy from $E = mc^2$) in order to maintain the average density of the Universe constant, since this would otherwise fall as it expanded. This could easily be calculated and the rate was so low as to be undetectable and therefore not disallowed by the many experiments testing the conservation of energy and mass.

The steady state theory sent major ripples through the astronomical community, and many senior, traditional astronomers reacted strongly against it, sometimes with personal overtones. In an address to the Royal Astronomical Society, a distinguished astronomer referred to it in a disparaging way by saying that he much preferred to call a spade a spade and not a perfect agricultural implement. But the theory had elegance and simplicity, and it also stimulated cosmological debate and cosmological observations because it was open to the test of observation. It proposed that the Universe did not change, on average, with time and, therefore, observations of distant objects, which are seen at earlier epochs, should show no statistical differences from nearby objects. The first test of this prediction was made by Martin Ryle, with the radio telescopes near Cambridge, which were carrying out sky surveys. Although the nature of the radio sources was not known, nor their distances, the prediction that the Universe was the same, on average, at distant epochs, led to a prediction of the relation between the number of radio sources and their intensity. The survey showed that there was a slight excess of the number of radio sources of lowest intensity. This implied that either the spatial density of distant radio sources was greater than locally or that the radio emission of the early radio sources was greater than now. In each case, the steady state theory was disproven. However, the deviation was rather small when compared to the scatter of the data, and supporters of the steady state theory were not unduly concerned. But then, in 1965, an unexpected and virtually accidental discovery was made, which was the most important cosmological observation since Hubble established the expanding Universe: the detection of the *cosmic microwave background*, which was to come down firmly and decisively in favour of the big bang cosmology.

The discovery was made by two radio scientists, Arno Penzias and Robert Wilson (no relation of mine) at the Bell Telephone Laboratories in New Jersey. They were working on a problem totally unrelated to astronomy: the efficient communication of information between a transmitter and a receiving antenna by radio waves of several centimetres in length. To do this, they had to track down the many sources of interference in order to eliminate them and thereby achieve noise-free transmission, but they reached a point where they still had a faint radio signal that they were unable to remove or identify. They investigated this background signal and found that it was always the same; it did not change over a day or through the seasons of a year; it was also the same wherever they pointed their antenna in the sky. The radiation had to be cosmic in origin, and measurements at other wavelengths showed it was consistent with Planck radiation at a temperature of about 3° Kelvin. They had detected the remnant radiation of the big bang.

Although Penzias and Wilson did not know it at the time of their discovery, the existence of a pure Planckian radiation field had been predicted previously as a consequence of the big bang, first by George Gamow and then by a group of astronomers at Princeton University, not far from the Bell Laboratories, who were actively setting up equipment to detect it. But they were beaten to it by two non-astronomers who stumbled onto it and were to be rewarded with a Nobel prize.

The explanation of this cosmic microwave background radiation field fell naturally and elegantly out of the big bang theory. In the early stages of the Universe, the matter would be very hot and very dense, and would be opaque to radiation. The material and light would therefore be in thermodynamic equilibrium with each other; the distribution of energy between the particles of matter would be given by Maxwell's law (Ch. 6) and the distribution of energy between the photons of light would be given by Planck's law (Ch. 7), each determined solely by the temperature. As the expansion proceeds, the temperature will drop until the medium ceases to be ionized, and the hydrogen and helium will recombine. At this point, the opacity to light, which was attributable to the free electrons, drops to near zero and the radiation immediately becomes uncoupled from the matter; it then spreads out at the velocity of light while the matter in the Universe continues to expand at a more sedate rate. The time at which the material and radiation in the Universe uncoupled is a few hundred thousand years after the initiation of the big bang, when the Universe was one ten-thousandth of its present age. We now see the event in the very distant past, but the Doppler shift caused by the expansion of the Universe has caused the intense fireball of 3000° Kelvin to appear as a faint microwave signal of about 3° Kelvin.

The cosmic microwave background represents the earliest observation that can be made of our Universe and it has therefore been studied intensively since its discovery. In particular, an American satellite, the *Cosmic Background Explorer* (COBE) has been devoted entirely to its study; launched in 1989, it is still operational at the time of writing. COBE has the great advantage of being able to observe the radiation over the whole of its spectrum and over the whole sky. Its continuous monitoring built up a measurement of unprecedented accuracy, which showed a perfect Planck distribution for a temperature just below 3° Kelvin. This is what the big bang theory predicted.

The cosmic microwave background radiation is not completely uniform over the whole sky but is slightly anisotropic, showing an increased temperature of one part in 1000 in one direction and a similar decrease in exactly the opposite direction. This is caused by a Doppler effect, which results from the motion of the Earth against the fixed background of the Universe, as indicated by the microwave background; if the motion of the Earth around the Sun and the motion of the Sun around the galactic centre are removed, it shows that our Milky Way galaxy is moving relative to the whole Universe in the direction of the constellation of Virgo at a speed of 600 kilometres per second. But if this galactic motion is removed in the analysis, it is found that the cosmic microwave background is extremely constant over the whole sky, with no sign of any fluctuations from one direction or another. This shows that the early Universe was remarkably uniform and it poses a major cosmological question as to how such a uniform beginning could have evolved into the highly structured Universe of stars, galaxies and clusters of galaxies that we see today. This problem prompted a detailed statistical analysis of the COBE data, which led in 1992 to the conclusion that there were fluctuations in the sky over angles of about 10° on the minute scale of one part in 100000. Recent ground-based investigations have succeeded in determining the actual structure of the background, but this cannot be related to the present-day structure because the fluctuations are seen in the remote past before they have had a chance to grow into stars and galaxies.

When Gamow first proposed the hot big bang cosmology in 1948, it was believed that the abundance of the elements was the same throughout the Universe, because this seemed to be the case for the Sun and the stars in the plane of the Milky Way. But as abundance studies extended to stars in the halo of our galaxy, such as those in the globular clusters, it was found that these had a different chemical distribution: they had very much less carbon, nitrogen, oxygen and all the other heavier elements, compared to hydrogen and helium. It was also realized that the halo stars had formed in the very early phase of the formation of the galaxy, and the disk stars had formed later (some, as was related in Ch. 11, very recently) and this showed that the abundance of the elements was different at earlier times than now. It was demonstrated by the work initiated by Hoyle and his collaborators that this was because of the creation of the heavier elements by nuclear transformation of hydrogen and helium in the interiors of the earliest stars, whose subsequent ejection into the interstellar medium enriched the stars that formed from it later. In our locality in the galactic disk, the abundances in terms of mass give hydrogen 74 per cent, helium 24 per cent, and all the rest 2 per cent. In the oldest stars of the galaxy, the heavier elements from carbon and upwards contribute only about 0.1–0.2 per cent of the mass. An extrapolation back in time implies that the primeval abundance that emerged from the big bang was of a composition entirely of hydrogen and helium, with a mass distribution of 75 and 25 per cent (about 90 per cent and 10 per cent respectively by number).

In about one second after the big bang, it is expected that matter was in the familiar form of protons, neutrons and electrons, plus neutrinos. It would also have the very high temperatures associated with the interiors of the hottest stars and, hence, nuclear synthesis processes will occur, starting with the fusion of hydrogen into helium, and if the high temperature and density were maintained, elements heavier than helium would also be processed, as in stellar interiors. But the temperature and density are decreasing rapidly because of the expansion, and thermonuclear processes cease after a few minutes when all the neutrons have been used up to give predicted primordial abundances by weight of 75 per cent hydrogen, 25 per cent helium and a tiny fraction of lithium and boron. Heavier elements were not formed and had to wait for stars to process them. A small amount of deuterium (the heavy isotope of hydrogen) is also predicted and its abundance is a very sensitive function of the density of matter at the time. The primeval abundance of deuterium is therefore a measure of the density of the Universe and, since this is a crucially important cosmological parameter, it will be returned to later. For the moment, it is enough to say that the prediction of the primordial composition of matter as being almost entirely hydrogen and helium (a prediction that is insensitive to the density of the Universe) is in agreement with the inferred result of the observations. As such, it was a second major success (after the cosmic microwave background) in favour of the big bang theory and it led most astronomers to accept that theory as the best basis for current cosmological thought. But wait awhile . . .

Another factor to be borne in mind in considering cosmological theories is the existence of intelligent life on Earth, and any model of the Universe must allow for this as well as the existence of stars, and so on. This was pointed out by the American astronomer Robert H. Dicke in 1961 and has become known as the *anthropic principle*, which simply says that the Universe has to be such as to allow the development of intelligent life. This places some constraints on the Universe: for example, its age must be about 10000 million years because that is the time it takes for the first stars to form, to build the elements heavier than hydrogen and helium that are essential to life and eject these into space where new stars (like the Sun) and planets (such as the Earth) can form, and the evolution of life can take place. It is remarkable that the Universe we observe and the laws of physics (and chemistry) seem to be just right for the development of intelligent life: for example, if the matter in the early Universe had been processed mainly into helium rather than hydrogen, the lifetimes of stars would have been much shorter than needed for the evolution of intelligent life. Also, quite minor changes in some of the fundamental physical constants would preclude life development.

The anthropic principle has become very popular with some cosmologists, who have developed and strengthened it considerably, to the extent that it is called the *strong anthropic principle* to delineate it from the *weak anthropic principle* described above. The stimulation of the *strong anthropic principle* came from the fact that the Universe we observe seems to be peculiarly well adapted to allow the development of life to an intelligent stage. It is argued that this could hardly happen by chance, and a religious explanation of creation by a supreme being is avoided by the following scenario: the Universe is infinite and contains a very large number of local universes (of which the one that we observe and exist in is one) and that these local universes vary greatly in their nature and, possibly, even in their physical laws. Very few will have the conditions needed to allow the development of intelligent life and it happens that our Universe is one of them, possibly the only one. We are here by chance.

The weak anthropic principle is not a theory, but is clearly a valid concept that has the value of placing constraints on cosmological theories. The strong anthropic principle is highly speculative: it has been developed to explain why our Universe is just right for us, but, since we cannot see anything beyond our own Universe, it is not possible to subject it to observational test, and it will not be referred to again in this book.

The in-depth studies of big bang cosmology that followed its general acceptance revealed major problems that still remain today, although some ingenious solutions have been formulated. One problem arises from the fact that the cosmic microwave background is so remarkably uniform, to one part in 100 000, over the entire sky. Yet this is a medium that has to evolve into a highly structured Universe of stars and galaxies. This uniformity problem is even more exacerbated by what is called the horizon problem, which points out that there can have been no communication or contact between different parts of the cosmic background over distances of more than a few hundred thousand light years, because the radiation was emitted at that time after the big bang. Hence, the fastest means of communication, light, placed a communication limit of a few hundred thousand light years, so there is no way that matter separated by distances greater than that limit could have had any connection or contact with each other. On the sky, this separation of a few hundred thousand light years is equivalent to about 20°. How, then, is there such complete uniformity over the whole sky?

Another problem is posed by the present expansion rate of the Universe and its age. If the potential energy of gravitation is greater than the kinetic energy of the Universe, it will ultimately stop expanding and will fall back on to itself in a *big crunch*. If, on the other hand, the kinetic energy of expansion is greater than the potential energy of gravitation, the Universe will continue expanding forever. Another way of putting this is to say that, if the ratio of these two energies is greater than unity, the

Universe will ultimately collapse; if it is less than unity, it will continue to expand. The gravitational energy depends on the average density of the Universe, and the kinetic energy on its velocity of expansion; in principle, therefore, these quantities are measurable and their values will be discussed later, but suffice it for the moment to say that limits were quickly established which placed the ratio between 10 per cent and 200 per cent. Now the average density of the Universe is decreasing because of its expansion, and the velocity of expansion is decreasing because of the gravitational retardation. Hence, both gravitational and kinetic energies are changing with time, and their ratio is changing with time. If at the initiation of the big bang the ratio is slightly greater than unity, it will continue to increase with time, resulting ultimately in a collapse; if the ratio is slightly less than unity, it will continue to decrease with time, resulting in a continuously expanding Universe. But the Universe is still expanding after 15000 million years (its age will be returned to shortly) and the restriction of the range of possible values of the energy ratio to 0.1 to 2 means that the initial value had to be very close to unity. How was the Universe created in this unique fashion, where the kinetic energy almost exactly balanced the gravitational energy? Or, as it is usually put, why is the density of matter in the Universe close to the value that will make it, in general relativity terms, flat and not curved?

It is now appropriate to go through the various phases of evolution of the Universe from its creation to its ultimate fate, as seen by big bang cosmology. The first phase covers the time from the beginning (zero) to about a millisecond and is undoubtedly the period of greatest uncertainty (some might say ignorance), which presents immense problems, not only in science but also in philosophy. Applying Einstein's theory of general relativity to a homogeneous expanding Universe allows its past temperature and density to be calculated and this predicts that, at the very beginning, the temperature and density were infinitely high. This comes from the mathematics and is called a singularity, but you should realize that, although singularities have occurred mathematically in other theories, nature has yet to reveal one unambiguously.

Astronomers realize that the nature and physics of the Universe in the instant immediately after its creation is beyond present knowledge and understanding if the theoretical extrapolation back in time is accepted. It is also known that general relativity will be invalid in that very early phase, because it is a continuum theory which will have to give way to quantum effects, just as the classical theory of electromagnetism had to give way to quantization in the early twentieth century. Also, the known forces of nature – the electromagnetic, nuclear and gravitational forces – will not be able to act separately under such extreme conditions and will appear in the form of a single force. This has led to searches for a grand unified theory (GUT) which will unify all the forces of nature, a problem that defied the strenuous efforts of Einstein in the latter half of his life. It is in this first instant after the birth of the Universe that the solutions to the major problems of uniformity, horizon and balanced Universe have to be found, but as we do not understand the conditions or the physics, any proposals are, of necessity, highly speculative.

One such proposal was made in 1981 by an American theoretician, Alan Guth, who realized that the difficulties could be overcome if there was a period of very rapid *accelerating* expansion in the early Universe, in contrast to the present slowly *decelerating* expansion. This scenario (it cannot be called a theory, because its basis rests on unproven physics) has been called *inflation* (a topical term coined from the economic problem of monetary inflation in the 1980s). It proposed that, in the first instant of the Universe, when all the forces we recognize now were merged into one, there was a *repulsive* force that inflated the Universe by an immense factor in a tiny fraction of a second. At the start of inflation, the densities were so high and the size of what has become our present Universe was so small that there was no horizon problem; everything was so compact as to be in close contact and in full equilibrium, of uniform density and temperature. Then, a phase transition occurs (like water into steam, but much more extreme), at the completion of which the electromagnetic, nuclear and gravitational forces will uncouple at a point when the system is in balance, that is, when the kinetic and gravitational energies are *exactly* equal.

The fact that the inflationary scenario can explain the apparently unexplainable problems of the big bang theory, which no other proposal can do, has led to it receiving considerable attention from cosmologists. But the reality of the postulated repulsive force is uncertain, although its existence cannot be ruled out by the current knowledge of particle physics; indeed, some of the future high-energy accelerators may throw some light on its nature. In order to solve the uniformity and horizon problems, the requirement on the magnitude of inflation is enormous; indeed, of all the scales and values that I have tried to communicate to you in some understandable way in this book, this is the most difficult by far. In relative terms only, it is equivalent to the inflation of a grain of sand to the size of the present observable Universe. It also requires an inflation rate that exceeds the velocity of light, which is essential to overcome the horizon problem.

The inflationary concept was able to explain the high uniformity in the Universe at the time when we see the cosmic microwave background, but, of course, some fluctuations had to be present in order to allow the growth of gravitational condensations to form the galaxies we see today. The absence of such fluctuations raised doubts, not only in the inflationary scenario, but in the big bang cosmology itself; but very recently, in the 1990s, tiny perturbations have been detected in the cosmic microwave background at the level of one part in 100000. Whether these are enough to produce today's stars and galaxies is a matter to be addressed shortly, but for now the question is whether the inflationary concept can explain such fluctuations. The strict answer is no, because the physics of the medium is not known, but a general answer of yes can be given on the following argument: although the conditions postulated would result in the highest degree of uniformity, there will be a very small but finite level of fluctuation attributable to quantum effects. Heisenberg's uncertainty principle rules out the possibility of total uniformity, and quantum fluctuations that would grow to the observed values of about one in 100000 seem a reasonable consequence.

The inflationary hypothesis is a scenario developed to explain certain observations and, in so doing, proposes some unproven scientific phenomena. However, it has the merit required of all scientific theories, of making a prediction that can be subjected to scientific test. It predicts that the Universe is *exactly* balanced, with its kinetic energy of expansion *exactly* equal to its potential energy of gravitation. In order to check this, the expansion rate of the Universe and its total mass (or average density) must be determined. The former is relatively easy, but the latter, beset by the problem of dark matter or missing mass, is very difficult, and this issue will be discussed a little later when the question of a closed or open Universe is addressed.

Another consequence of the inflationary Universe or, perhaps, any other big bang concept, is that, if the initial Universe was infinite or at least much larger than the total material forming our own observable Universe, then the possibility of other universes exists. But these would be beyond our horizon and undetectable.

During the inflationary period, it is proposed that there was no matter, but only energy in the form of the single unified force. When the inflation is complete, the single force will break down into the electromagnetic, nuclear and gravitational forces we experience today, and the energy will convert into matter. This exposes another major problem in big bang cosmology (and probably in any other cosmology as well). Experiments in high-energy particle physics, in which energy transforms into matter, show that all processes result in the formation of two particles: a particle and its antiparticle (see Ch. 7). For example, given that it has sufficient energy, a gamma-ray photon can transform into a proton and an anti-proton, or an electron and a positron, and this *symmetry* has been strictly maintained in every single case. Hence, if such rigorous symmetry in the transformation of energy into matter applies in the early Universe, then in the post-inflationary period, the medium had to be composed of an identical number of particles and anti-particles. But the Universe we see today is a matter Universe, not an anti-matter Universe. We know this in several ways, of which the most direct is the fact that all material striking the Earth from outside is in the form of matter, including the cosmic rays that come from our own galaxy and from external galaxies. The only way to explain a matter Universe is to postulate a breakdown in the symmetry principle at early times, so that there is an excess of matter over anti-matter; the excess required is rather small, being only one part in a thousand million.

In the immediate post-inflationary period, the medium consists mainly of the elementary particles and their anti-particles. These do not include protons and neutrons, which are not elementary, but do include electrons and neutrinos, which are. It is now generally accepted that the basic constituents of protons and neutrons are smaller elementary particles, which have been called *quarks*, after a term used by James Joyce in his book *Finnegans Wake*, in which one of his characters issued the somewhat enigmatic call "Three quarks for Muster Mark". There are six different quarks, which have charges of one-third or two-thirds the charges of the proton (positive) or the electron (negative). Different combinations of these form the proton (charge plus one) or the neutron (charge zero), in which they are highly bound. There are also six anti-quarks, which can form an anti-proton or an anti-neutron. Since particles and anti-particles have a strong affinity for each other, they will quickly combine by annihilating each other and producing two high-energy gamma-ray photons. If complete symmetry had existed in the creation of matter and anti-matter, the Universe would have become one composed entirely of energy in the form of gamma rays, but the postulated slight excess of the former over the latter will leave some matter particles without an anti-matter mate, and these will form into protons, neutrons, electrons and neutrinos. These will be in full equilibrium with each other and with the much more numerous (by a thousand million) gamma-ray photons. It was therefore a radiation dominated Universe.

The initial instant of the Universe, as seen by big bang cosmology, is the most uncertain period of all, because the basic science is not known and this has resulted in unproven, albeit plausible, hypotheses being proposed in order to explain the present Universe. But after one millisecond, the extrapolation back in time shows that the temperature will have declined to about a million million degrees Kelvin and the conditions would be such that the knowledge of existing physics should apply. Also, since the conditions prevailing at that time are relatively simple, with radiation and matter in full thermal equilibrium, the period from then until a few hundred thousand years is one in which big bang cosmology has the greatest confidence and one from which its predictions have been the most successful.

Up to about one second, even the highly un-interacting neutrinos are confined by the very high-density medium, but at that point the density, although still high by any other criterion, has decreased to a level that allows them to decouple and escape. Up to that point, the equilibrium between the constituent particles was such as to maintain closely equal numbers of protons and neutrons, since they were equally easy to produce in interactions involving neutrinos. When these escaped, the production rates dropped dramatically, causing the situation to change markedly because the proton is a stable particle, whereas the neutron is not, decaying into a proton, electron and an anti-neutrino with a half life of about 15 minutes. This process reduced the number of neutrons (and increased the number of protons) and, if this had continued, the primordial composition of the Universe would have been entirely hydrogen. But nucleosynthesis processes started when protons were seven times more numerous than neutrons and these were mopped up into helium during the first few minutes, thereby explaining the primordial abundance of the elements of 75 per cent hydrogen and 25 per cent helium by mass.

After a few hundred thousand years, the temperature of the Universe had dropped to about 3000° Kelvin, the hydrogen and helium became neutral, thereby becoming transparent to radiation which, no longer being confined by the medium, escaped, to be observed as the remnant cosmic microwave background. This, together with the primordial abundances, established the big bang as the leading cosmological theory.

The next phase to be considered in the evolution of the Universe in terms of big bang cosmology is that from the uncoupling of radiation and matter after a few hundred thousand years to the present day, a period of about 15 000 million years. At the start of this phase, the Universe was extremely uniform, as shown by the cosmic background radiation, and now it is highly structured in the form of massive aggregates of stars called galaxies, which are themselves in clusters and superclusters. How this happened is one of the most challenging current questions in astronomy. The general answer is gravitational contraction, but the Universe is not static but expanding, and this requires a clumping in the matter sufficient to produce a local gravitational field which can overcome the kinetic energy of expansion. Calculations show that this requires perturbations in mass of about one part in a thousand over regions that embrace sufficient mass to form galaxies, most of which contain matter equivalent to between 100 million and 1 000 000 million suns. But the fluctuations detected in the cosmic microwave background are only one part in 100 000 and, since matter and radiation were in complete thermal equilibrium with each other at that time, this means that the fluctuations in mass were also only one part in 100 000; this quantifies the problem.

However, we know that stars and galaxies exist and we believe that there is no other possible mechanism for their formation than gravitational contraction. This has led to the proposition that there is some other form of matter in the early Universe which is not coupled to the radiation and is therefore not in equilibrium with it; consequently, it need not be uniform and could be clumpy, depending of course on its nature and properties. An obvious candidate is the neutrino, as these particles uncouple from matter and radiation when the Universe is only one second old. But these particles would have to have a significant rest mass (i.e. there must be some mass left after their absorption) and all that is known at present is that this lies between zero (like the photon) and 5 electron volts (remember that particle physicists always give mass in terms of energy via $E = mc^2$). It is therefore a very light particle (the rest mass of the electron is about half a million electron volts and of the proton a thousand million electron volts) and hence a very speedy one, which means that any clumpiness would happen only over very large distance, too large for galaxy formation.

Another possible candidate for producing sufficient gravitational perturbations to initiate galaxy formation is dark matter. This was discussed in the previous chapter, when it was revealed that the visible matter we see in galaxies is only 10 per cent of the total mass. Ninety per cent is in the form of some dark matter, of which we know nothing except that it exerts a gravitational force. The inflationary scenario for the first instant of the Universe predicts even more dark matter, reducing visible matter to only 1 per cent of the total. The question of dark matter will be discussed again shortly in the light of the future evolution of the Universe. For now it is sufficient to say that 90 per cent of the Universe is in the form of some kind of dark matter and this possibly could be 99 per cent. So, if this dark matter was not coupled to ordinary matter and radiation at the time of the release of the cosmic background radiation, it is possible that it could be clumpy and produce the required gravitational perturbation of one in a thousand over the necessary dimensions. But the perceptive reader will recognize that this possibility is based on ignorance; we need knowledge in order to rule it in or out.

If, indeed, the early gravitational perturbations are sufficient to allow galaxy formation, the actual details of that process are also far from being understood. Although there is a good understanding of stars, their types, nature and evolution, there is no corresponding picture for galaxies. These vary enormously in their mass, size, shape, nature and stellar content. Most are spiral rotating systems, but there are also elliptical and irregular galaxies; further, in addition to these "normal" galaxies there are the active galaxies, of which the most energetic are the quasars. How these form and then evolve is a basic astronomical problem exacerbated by two factors. First, galaxies do not evolve as isolated systems but are subject to collisions with other galaxies, which cause major changes in their nature whether they merge into one system or not. As an example, let us take our own Milky Way galaxy. Since its formation, this has accreted one-third of the matter now present in it, and in 500 million years it will collide with the newly discovered dwarf galaxy that lies about 50 000 light years away on the other side of our Milky Way and has been obscured by it. In about 5000 million years we will merge with the Large Magellanic Cloud and in about 10 000 million years we will probably collide with the Andromeda spiral galaxy.

The second factor is an observational one. In looking at fainter and fainter galaxies, we are looking further and further back in time, so, in principle, we should be able to trace out the evolution of galaxies. However, their formation occurs when the Universe is in the first 10 per cent of its present age and, at such great distances, galaxies are too faint to be observed. Hence, direct evidence of the formation process is not currently available. This, as much as any other factor, has led to the current construction of super-large optical telescopes in the Northern and Southern Hemispheres. It may be that a breakthrough in this problem will have to wait for their full completion and operation.

Although we can look at the Universe back into the past, it is not possible to see into the future, and how it develops from now on is a matter for prediction. The most basic question to consider is whether the Universe has sufficient kinetic energy to continue expanding, because, if it has not, gravity will ultimately stop it and it will fall back on itself in a big crunch. To decide on these two options we have to measure the rate of expansion of the Universe and the total mass within it.

The rate of expansion was first estimated by Hubble in 1929 but, since it required a measurement of one of the most difficult quantities in astronomy, distance, it was a very inaccurate estimate. Extensive investigations in the post-war years have greatly improved the accuracy to the point where it is generally accepted that it is known to within a factor of two and, as some believe, including me, to about 10 per cent. The standard methods have used the observation of objects whose intrinsic luminosity is believed to be known and, from their measured brightness, a distance can be determined. For cosmological purposes, these standard objects have to be very bright and they include the cepheid variables, but other objects such as supernovae at peak brightness have also been used. These methods have tended to give a range of values over a factor of two and, of necessity, are constrained to the relatively local Universe, within about 10 per cent of the diameter of the whole Universe. Rather than give the values in the conventional astronomical units, I will express them in the more easily understood form of a lifetime; that is, the time from the start of the Universe to the present era if the expansion velocity had remained constant. This is not the same as the age of the Universe (but is related to it), as will be discussed shortly. This "Hubble" time varied between 10000 million and 20000 million years, as derived by the conventional methods. A quite different method, and one extending to a far greater distance, was afforded by the discovery of the double quasar reported in the previous chapter. This is a double image of a single quasar formed by an intervening galaxy acting as a gravitational lens and, as the light paths in the two images are not equal, the time difference that light takes to travel along each of them is directly related to the Hubble time. Variability in the ultraviolet emission from this quasar enabled this time difference to be measured by the IUE satellite and gave a Hubble time of 15000 million years. This is the value that is adopted in this book.

The other parameter to be measured in order to assess whether the Universe is open (expands forever) or closed (will contract) is the total mass of the Universe. There is a critical mass below which the Universe is open and above which it is closed, and estimates will be presented in terms of that critical mass. The Universe that we can see in the stars and galaxies is 1 per cent of the critical mass but, as reported in Chapter 12, there is much more dark matter associated with the galaxies, which has been revealed by its gravitational effect. This places the mass of the Universe at 10 per cent of the critical mass. However, the inflationary scenario of the early Universe requires that the Universe will have the critical mass precisely. From another source (the present dynamics of the Universe and the fact that it is still expanding after about 10000 million years) an upper limit to the mass can be established as twice the critical mass. If this was the true mass, the Universe would stop expanding and reverse into a big crunch in 50000 million years, or about four to five times its present age.

Hence, at the present time, estimates of the mass of the Universe vary between 10 per cent and 200 per cent of the critical mass, so, as far as we know, the Universe can be open or closed. But the estimates do have different degrees of confidence. The value of 10 per cent of critical is based on very well established principles and can be regarded as a firm lower limit. It is also backed up by a completely different argument, which comes from the theory of the thermonuclear processing of the elements in the first few minutes of the big bang. This correctly deduced the primordial abundance of the elements as being predominantly hydrogen and helium, but with minute traces of deuterium (heavy hydrogen) and lithium. But, unlike hydrogen and helium, the amount of deuterium depends very sensitively on the density of the medium and therefore of the mass of the Universe. This is because deuterium is easily destroyed and fused into helium, causing it to be greatly reduced at higher densities than at lower densities. The abundance of deuterium in the local interstellar medium has been measured by ultraviolet observations and gives a value of two parts in 100000 in agreement with a Universe of 10 per cent of the critical mass.

Another factor in favour of the value of 10 per cent comes from considerations of the age of the Universe. This has to be less than the Hubble time and, if the Universe was exactly balanced, its age would be two-thirds of that time, or 10000 million years. However, estimates of the ages of the globular clusters in our own galaxy made by charting the evolution of the stars within them (Ch. 11) give values up to 14000 million years. These are the oldest stars, which formed when the galaxy formed and therefore give an estimate of the age of the galaxy that is older than that deduced for a balanced Universe. On the other hand, the age for a Universe with 10 per cent of the critical mass is consistent, within the errors, with that derived for the oldest stars.

The firmest estimate of the mass of the Universe is therefore 10 per cent of critical, but this would make the inflationary scenario untenable and, if this were so, the major problems it was constructed to solve – of uniformity, balance and horizons – return. At the present time we do not have sufficient confidence in the estimates to say that the Universe is open or closed, but the probability lies with the former.

If the Universe is open, it will expand forever (but at a decelerating rate because of gravity) and the matter within it will run down as the second law of thermodynamics takes its toll. The more massive stars (greater than about eight solar masses) will evolve most quickly and end their lives as cold neutron stars, or possibly black holes if there are some very long-term gravitational effects. The less massive stars will evolve more slowly to become white dwarfs and ultimately black dwarfs as they cool down. The least massive of stars have about one-tenth of a solar mass and evolve extremely slowly, taking about 10000000 million years to reach their final dead state; this is about a thousand times longer than the present age of the Universe.

If the Universe is closed, a totally different future lies ahead. The expansion will stop and it will start to contract with ever-increasing velocities.

This will happen in 50 000 million years at the earliest, when the least massive stars will still be shining together with any recently formed massive stars. The radiation from these will now be blue-shifted and, as the contraction speeds up, the radiation temperature will rise until it exceeds the temperatures of stellar atmospheres and, ultimately, of stellar interiors. Thermonuclear processes will be greatly accelerated everywhere and finally there will be a great crunch and fireball, but this will be very different from the initial big bang since it will be very lumpy and composed mainly of the heavier elements.

The above predictions for the future of the Universe are made on the basis of the matter and objects we can see, and you should remember that this accounts for only 10 per cent of the Universe and, if the inflationary scenario is accepted, only 1 per cent. Hence, the future may be conditioned more by the dark matter we cannot see than by the matter we can see. But we know nothing about the nature of the dark matter, other than that it exerts a gravitational field. Consequently, there has been a wide range of proposals from the mundane to the exotic. The mundane proposals centre on bodies composed of familiar matter, but insufficiently massive to have ignited into a star, together with dead stars that have completed their life terms. The astronomical evidence in favour of these is rather weak. The exotic proposals centre on the possible existence of a totally different kind of matter, one not composed of the familiar protons, neutrons and electrons that everything, including ourselves, is made of, but something else. Since we have not detected such matter, it must be weakly interacting and have mass. Its basic particles have therefore been dubbed WIMPS – weakly interacting massive particles – and devices have been placed in deep mines in the hope of detecting them. The chances seem rather remote, but the reward for a positive detection would be immense.

In summary, big bang cosmology can explain the expansion and the early composition of the Universe, together with the cosmic microwave background – three major successes that have led to its general acceptance by most astronomers. But, as discussed above, extremely severe problems remain, particularly concerning the very first instant, when the conditions are outside the scope of any human experience or experiment and therefore are not understood. An ingenious proposal to solve these problems, such as the extreme uniformity of the initial Universe, is called inflation and it requires a tremendous repulsive force which initiates the expansion and causes velocities greater even than that of light. But this scenario also predicts a balanced Universe, which leads to an age determination shorter than that obtained for the oldest stars in the Milky Way galaxy. But possibly the most puzzling aspect of cosmology is the nature of dark matter and the role it plays. We are confident that 90 per cent of the Universe is composed of unseen matter and, if the inflationary hypothesis is correct, this becomes 99 per cent. The Universe is therefore composed mainly of some kind of matter, whose nature and properties are completely unknown, but we have no choice but to base our theories of cosmology on that part which we can see and understand.

It is appropriate to end by repeating a warning given above. We can observe the Universe only as far back as the cosmic microwave background, when the radiation uncoupled from matter and escaped; this happened when the Universe was a few hundred thousand years old, or a ten-thousandth of its present age. Before that, we must rely on theory – particularly general relativity theory – to predict densities and temperatures. The success in explaining the primordial abundances of the elements gives considerable confidence in this theory as far back as when the Universe was one millisecond old. Before that, the densities and temperatures that emerge take us into a realm that lies outside our experience and beyond our knowledge. Further, the mathematics of our present theories contains a singularity where the densities and temperatures become infinite at a time of zero, when there has to be instantaneous creation. Before that, there is no matter, no energy, no space and no time. However, although singularities have occurred before in the mathematics of some scientific theories, nature has yet to reveal one in practice. But perhaps the creation of the Universe will prove to be the exception.

Epilogue

What is the nature of the Universe?
What is our place in it and where did it and we come from?
Why is it the way it is?
. . .
If we find the answer to [these questions] it would be the ultimate
triumph of human reason for then we would know the mind of God.
Stephen Hawking

We have come a long, long way since the very first curiosity and mystically driven studies of the heavens. The sedate, static Universe of the ancient Greeks, consisting of planets and stars stretching out to just beyond the orbit of Saturn, has been replaced by a violent, dynamic Universe extending out to an unimaginable distance of 10 000 million light years. To the relatively gentle stars have been added the remarkably energetic quasars and X-ray binaries, with their relativistic gravitational power generators. The super-dense neutron stars have been revealed as pulsars and have been shown to be the end product of supernova events, the most immense of cosmic explosions. Further, the broad characteristics of the observable Universe as far as its outermost limits have been established, enabling a serious attack to be made on the great question of the origin and evolution of the Universe.

However, with these tremendous advances, it is as well to pause at this point and avoid the error of believing that all is now known, because we are still only tapping the foundations of our knowledge about the Universe. To press this point home, I have to stress that, despite the great advances in understanding of many of the various objects in the Universe, there are many phenomena of which we are totally ignorant. These include the gamma-ray bursters (Ch. 12) whose location and nature are completely unknown and whose mechanism of intense, high-energy power generation can only be guessed at. Even more puzzling is the nature of the dark, unseen matter which has been detected only by its gravitational effects. Since this comprises the bulk of the matter in the Universe and we have no idea of its composition, it clearly raises a major question mark over theories of the Universe, since these are based mainly on the matter we can observe directly, and this is only a small fraction of the total.

The big bang theory of the origin and evolution of the Universe is the one that has current general acceptance, but it does have some major problems, which have been outlined above and which will be reviewed again now. In this case I will go backwards in time from the present local Universe to the distant early Universe. We see a complex structure of stars, galaxies and clusters of galaxies which has had to evolve from the very uniform and smooth early Universe, as seen in the cosmic microwave background radiation. How this evolution occurred is one of the major current interests in astronomy, and the complexity of the problem has so far denied a full solution. But we do see the Universe in these quite different states at quite different times, so the evolution from one state to another must have happened and this by the only possible process of gravitational condensation.

From the age of a few hundred thousand years when the radiation escaped, which we now see as the cosmic microwave background, back in time to less than a second, the Universe was in its best understood phase according to our current knowledge of physics. In the first few minutes, thermonuclear processing predicts the production of matter composed almost entirely of hydrogen (75 per cent by mass) and helium (25 per cent). This is in very good agreement with the primordial abundance of the elements deduced from observations of the oldest stars; this fact, together with the successful explanation of the cosmic microwave background radiation, led to the general acceptance of the big bang theory of cosmology.

However, as we go to earlier times, particularly to an age of less than a millisecond, the predicted values of density and temperature take us into a realm that is far beyond our current scientific knowledge; the one thing we can be certain about is that our current theories of physics, including Einstein's theory of general relativity, are invalid. And yet it is in this first instant that the entire nature of the subsequent Universe was established. Questions such as the primordial abundance of the elements and the cosmic microwave background were answered in a later phase, but other major questions can be answered only by the processes occurring in the very first instant.

One of these major questions is why the Universe, as observed by the cosmic microwave background radiation when it was a few hundred thousand years old, is so very uniform, even though the matter and radiation

278

in it are separated by distances greater than can be travelled by light in that time. Hence, there can have been no communication (the horizon problem), never mind the direct physical contact that would be needed to establish uniformity by equilibrium. An attempt to solve this problem led to the proposition of an inflationary period in the first instant of the Universe, when an immense repulsive force is invoked to expand a super-dense medium, which is in close contact and therefore in full equilibrium, into a new phase. The separate parts are now no longer in contact because they are separated by more than the light-travel times, but they retain the memory of the earlier phase and are therefore in full equilibrium. Clearly, to do this requires an expansion rate greater than the velocity of light, in contradiction to Einstein's special theory of relativity.

Another major problem that arises from considerations of the first instant of the big bang is why a matter rather than an anti-matter Universe emerged from the initial creation of matter. All experiments in particle physics show that the transformation of energy (usually in the form of high-energy gamma rays) results in a pair of particles – a matter particle and its anti-matter component. But we live in a Universe composed of matter, which implies some breakdown of symmetry in the very early Universe. (Recent experiments in high-energy physics show that in some particle reactions complete symmetry is not maintained.)

To the above major problems can be added an even greater one, which is as much philosophical as it is scientific. An extrapolation back in time, using the mathematics of the big bang theory, results in a starting point where the Universe was infinitely dense and infinitely hot, but before which there was nothing, no matter, no energy, no space and no time. The Universe had to appear in an instantaneous flash; there had to be a creation out of nothing. Based on our current knowledge and understanding, there is no possible scientific explanation of this and some readers may fall back on a religious explanation, that is that a supreme entity created the Universe some 15000 million years ago and established its nature and its laws. There is nothing in the scientific study of the Universe to disprove such a proposition but neither is there anything to support it. The question of a supreme entity is quite beyond the current level of scientific study and may remain so into the indefinite future. Since the Universe we can see and detect is still beyond a full scientific explanation, the possibility of understanding its creator, should there be one, is so remote as to be discarded. Nevertheless, the advances that the human race has made in understanding the Universe have been immense but, this notwithstanding, it is quite likely that, in the not too distant future, learned human society will look back at our present construction of the Universe in the same way as we now look back on that of Aristotle.

Glossary

aberration   The apparent shift in the position of a star viewed by a telescope because of the motion of the Earth across the direction of the starlight. Its magnitude is determined by the Earth's velocity and the speed of light, and has a maximum value of 20 seconds of arc. (Aberration is a term also used to describe distortions of light beams in optical instruments, but it is the astronomical meaning, described above, that is used in this book.)

absorption lines   Dark lines in a spectrum (like that of the Sun) caused by the absorption of atoms in the line of sight at specific wavelengths.

accretion disk   A disk of material spiralling around a body as it falls into its gravitational field, generating heat by friction as it does so.

active galaxy   A galaxy which is generating an immense level of power because of a giant black hole in its nucleus, which sucks in matter (gas, dust, stars), thereby releasing gravitational energy. They include quasars, Seyfert galaxies and radio galaxies.

alpha (α) particle   A particle consisting of two protons and two neutrons (mass 4, charge +2): it is the nucleus of the helium atom.

angular diameter   The apparent size of an object measured as an angle: the angular diameters of the Sun and Moon, as seen from the Earth, are both ½°.

antimatter   The ordinary matter with which we are familiar is not the only form of matter. There is another form called antimatter which has the same mass but is opposite in every other sense such as electric charge. At the particle level, the antiparticle of the electron is a positron, for the proton it is an antiproton, and for the neutron it is an antineutron. When matter and antimatter come into contact, they annihilate each other to produce pure energy, usually in the form of gamma rays, according to Einstein's formula $E = mc^2$.

aphelion   The point in the orbit of any body in the solar system when it is farthest from the Sun.

asteroids   Small rocky bodies orbiting the Sun and lying mainly between the orbits of Mars and Jupiter; believed to be the debris of the break-up of planetary-size bodies early in the formation of the Solar System.

astronomical unit (AU): A unit of distance equal to the average distance of the Earth from the Sun. It is equal to 150 million kilometres.

astrophysics   The application of the laws of physics to the study of astronomy, mainly by the use of **spectroscopy**.

atom   The smallest chemical unit which can combine with other atoms to form molecules. The lightest is hydrogen (H) and, if two of these combine with an oxygen atom (O), the resulting molecule (H_2O) is water.

atomic number   This defines the chemical properties of each atom (element) and is equal to the number of protons in its nucleus. The atomic number of hydrogen is 1 and of helium 2.

atomic weight   This is the mass of the nucleus of an atom and is equal to the number of protons plus the number of neutrons. The latter can vary for different isotopes, but the chemical properties, caused by the number of protons, do not change. The atomic weight of the lightest element, hydrogen, is 1, but it has a heavier isotope (heavy hydrogen) of mass 2 (1 proton plus 1 neutron) – **deuterium**.

beta (β) particle   One of the three particles emitted by radioactive sources such as radium: now known to be an **electron**.

big bang theory   The cosmological theory by which the Universe started in a single, instantaneous event some 15 thousand million years ago and has been expanding ever since.

binary star   Two stars that are gravitationally trapped and in orbit about each other.

black body   A body that absorbs radiation of all wavelengths that fall on it and, hence, can emit all wavelengths as it heats up. It is a hypothetical body, which was used by Planck, together with his quantum hypothesis, to derive the distribution of radiation emitted by hot bodies. This distribution is commonly referred to as either **black body radiation** or **Planckian radiation**.

black dwarf   The very final stage in the evolution of stars such as the Sun. All energy sources have been exhausted leaving the star as a cold, dark body.

black hole   A body, or group of bodies, so compact that the gravitational field has become dominant to the extent that even light cannot escape. Originally a theoretical concept, they have now developed some reality because of the astronomical evidence of their existence in the nuclei of active galaxies such as quasars.

brown dwarf   A gravitationally collapsed object which is insufficiently massive to heat its interior to the required temperature (10 million °K) to ignite thermonuclear reactions and hence form a star.

cepheid variable   A class of variable star which pulsates and causes regular fluctuation in its brightness. The period of pulsation is correlated with the star's luminosity and this has made it a very important tool for determining luminosity and therefore distance.

chromosphere   The thin atmosphere just above the Sun's **photosphere** and below the Sun's **corona** which is heated by some form of mechanical energy deposition (possibly acoustic waves) plus conduction from the very hot corona.

comets   Minor bodies in the Solar System that have very eccentric orbits. Halley's comet is a very extended ellipse with a period of 76 years. They are believed to fall into the Solar System from outside and may escape after orbiting the Sun, but if they are perturbed by any planets, they can become trapped like Halley.

constellation   A pattern of bright stars in a region of the sky which is often associated with some mythological figure: an example is Orion, the Hunter, which represented the great god Osiris to the ancient Egyptians.

corona   The very extensive, very hot medium surrounding the Sun, which emits the X-rays that cause the Earth's **ionosphere**. Its heating is almost certainly attributable to mechanical energy transmitted from the dynamic motions in the Sun's interior by some form of propagating wave, whose exact nature is not understood.

cosmic background radiation   A radiation field that pervades the whole Universe and has the characteristics of a **black body** of temperature 2.7°K. It is the remnant of the very hot radiation field of the initial big bang.

cosmic rays   Extremely energetic particles travelling at velocities close to that of light, which bombard the Earth's atmosphere from all directions. They are composed of positively charged atomic nuclei, of which the most abundant is that of hydrogen, the **proton**, together with negatively charged **electrons**. Being charged, the rays are bent by the magnetic fields in the galaxy and their origin, unlike light rays, cannot be deduced from their direction

cosmology   The study of the universe as a whole including its origin, evolution and large-scale structure.

dark matter   Matter that is known to exist because of its gravitational effect but cannot be detected in any other way: its nature is therefore completely unknown. Dark matter comprises 90 per cent of the universe but current theories of cosmology require it to be at an even higher level of 99 per cent.

degenerate matter   Matter that has become so compressed that the separation of the individual particles is less than the size of an atom (causing atomic degeneracy, as exists in white dwarfs) or less than the size of an atomic nucleus (nuclear degeneracy, as occurs in neutron stars). In these states, the pressure of the medium is determined by quantum effects and is higher and quite different from a normal gaseous state.

deuterium   The heavy isotope of hydrogen whose nucleus comprises a **proton** and a **neutron**.

diffraction   The degree to which light spreads out in passing through an aperture or around an obstacle. It is caused by the wave nature of light and is smaller for short wavelengths and larger for long wavelengths.

diurnal   Any effects caused by the daily rotation of the Earth, such as the rising and setting of the Sun and stars.

Doppler effect   The change in the wavelength of light caused by the motion of a source relative to an observer. If the source is approaching, it causes the wavelength to decrease (blueshift) and if it is receding, it causes the wavelength to increase (redshift).

$E = mc^2$ The equivalence of mass (m) and energy (E) linked by the velocity of light (c) in the expression derived by Einstein in his special theory of relativity.

eclipse   The blocking of all (total) or part (partial) of the light from one astronomical body by another in the line of sight. A total solar eclipse (by the Moon) is the best known and most spectacular example.

eclipsing binary   A double star system in which the two stars are periodically obscured by the other as they orbit each other in a plane which lies in the line of sight.

ecliptic   The plane of the Earth's orbit about the Sun. With the exception of Pluto, the other planets also orbit close to that plane.

electromagnetic radiation   Radiation caused by the propagation of waves of oscillating electric and magnetic fields. Depending on the wavelength (or frequency), these are known as radio, infrared, visible, ultraviolet, X-ray or gamma rays. They all propagate at the same velocity (*in vacuo*) – the velocity of light.

electron   An elementary particle that is negatively charged (–1) and which is very light (1/1830) compared to the proton. They orbit the positively charged nuclei of atoms and determine the chemical properties. Electrons are strong absorbers and emitters of electromagnetic radiation.

electron volt   A unit of energy that is equal to the energy acquired by an electron when accelerated by a electrical potential of one volt. It is denoted by eV and there are also units of a thousand eV (keV) and a million eV (MeV). The rest-masses of particles are usually given in terms of energy rather than mass (via $E = mc^2$); that of the electron is about 0.5 MeV.

emission lines   Bright lines in the spectrum of an object which are emitting more powerfully than the background source because of the high excitation of the atoms (and ions) responsible.

entropy   An indication of the amount of energy in a system which is unavailable to do useful work. It is also a measure of the degree of disorder in the system.

According to the second law of thermodynamics, entropy increases with time until it maximizes.

equinox The two times in a year when the Earth's Equator coincides with the plane of the Earth's orbit (the ecliptic). At those times, the duration of night and day are equal everywhere on Earth (except for some small differences caused by atmospheric refraction). The two times occur in spring (the vernal equinox) and autumn (the autumnal equinox).

escape velocity The velocity than an object must reach in order to escape from the gravitational field of the body on which it is located. For the Earth, the escape velocity is 11 kilometres per second, and for the Sun, it is 618 kilometres per second.

exclusion principle The principle developed by Wolfgang Pauli, which states that only one electron can occupy a single quantum state at any one time. This was instrumental in explaining the chemical properties of the elements and the pressure inside white dwarfs and neutron stars.

extragalactic astronomy The study of matter and galaxies that lie beyond our own Milky Way.

fission The splitting of a heavy atomic nucleus (e.g. uranium, plutonium), thereby releasing large amounts of nuclear energy. This is the process on which the first two atomic bombs was based and became the "trigger" for the hydrogen fusion bomb.

flare An explosive outburst from a localized region of the Sun's surface, which causes high local heating and the emission of high energy particles and X-rays.

fusion The binding together of light atomic nuclei (e.g. hydrogen into helium), thereby releasing very large amounts of nuclear energy. This is the main source of energy that powers the stars.

gamma rays Electromagnetic radiation of extremely short wavelength and extremely high frequency. They are the most energetic form known of electromagnetic waves.

general theory of relativity Einstein's theory of gravity, which replaced that of Newton in 1919.

globular clusters Spherically symmetrical clusters of stars located in the halo of the Milky Way galaxy. They contain up to a few hundred thousand members and are the oldest stars in our galaxy.

gnomon The first and simplest astronomical instrument, consisting of a vertical pole by which the altitude and direction of the Sun could be measured from the length and direction of its shadow.

gravitational lens Light is bent by a gravitational field so that a massive body, such as a galaxy, can deviate the light rays from a body that is seen behind it. This can cause a lensing effect, which makes the more distant body appear multiple. The double quasar was the first example.

gravitational red shift As light escapes from an astronomical body, it loses energy because of the gravitational field; this results in a reduction in frequency and hence an increase in wavelength – a red shift.

gravitational waves Waves in the form of an oscillating gravitational field, which travel at the velocity of light. They were predicted by Einstein's theory of general relativity but, as yet, they have not been detected directly.

greenhouse effect An effect in a planetary atmosphere, which causes additional heating and a temperature rise at the surface. The incoming solar radiation at visible wavelengths passes more easily through the atmosphere than the outgoing radiation, which is at infrared wavelengths and is heavily absorbed by molecules such as carbon dioxide. The atmosphere therefore acts as a blanket. The Earth has a greenhouse effect, as you can easily discover by going up a mountain, but the most extreme greenhouse effect is on Venus, which causes a temperature of about 700° Kelvin.

Gregorian calendar   This is the calendar we use today, with 365 days and leap years, which keep it in step with the seasons. It was instituted in 1582 by Pope Gregory XIII and was a revised form of the earlier Julian calendar (initiated by Julius Caesar), which was based on the early Egyptian calendar.

Heisenberg uncertainty principle   See uncertainty principle.

heliacal rising   That time of year when a star "reappears" after a period when it could not be seen because it was in daylight. It is seen again when it rises sufficiently ahead of the Sun to be viewable in the eastern sky just before dawn. The heliacal rising of Sirius had a special meaning in ancient Egyptian mythology as a herald of the flooding of the Nile.

Hubble constant   The present, local rate of expansion of the Universe. Its inverse is a time that represents the maximum possible age of the Universe, currently with a value of about 15 thousand million years.

inflationary universe   A cosmological theory concerning the very first instant of the Universe. It proposes a major change of state in the medium, which causes it to "inflate", almost instantaneously, from a highly compact homogeneous condition state into the kind of state expected for the start of the big bang.

infrared radiation   Electromagnetic radiation at wavelengths longer than the red end of the visible spectrum, and extending to about 1 millimetre, beyond which lies the radio-wave region.

intercalary   A period of time (usually a day, but in ancient times could be a month) that is inserted into a calendar to harmonize it with the seasons: for example, 29 February in leap years.

interference   The property whereby two coherent light beams interact when they cross. Bright fringes occur where the electromagnetic oscillations are in phase, causing them to reinforce, and dark fringes occur where they are in anti-phase, causing destruction.

intergalactic medium   The medium that lies in between the galaxies. So far, only clouds of pure hydrogen have been detected, but if, as believed, these are clouds of the original primordial material, they will also contain helium.

interstellar medium   The medium in between the stars which comprises both gas and dust

ion   An atom which has been stripped of at least one of its electrons. If hydrogen is ionized, it becomes a **proton** and a separate **electron**.

ionosphere   The upper part of the Earth's atmosphere, which is ionized by the Sun's ultraviolet and X-ray radiation.

Io torus   A ring of material surrounding the planet Jupiter and lying around the orbit of its nearest moon, Io. The material is mainly in the form of oxygen and sulphur, which has been ejected by Io's volcanoes, ionized by the Sun's radiation and then accelerated and partially trapped by Jupiter's rapidly rotating magnetic field.

isotope   Any element has the same chemical properties determined by the number of protons in its nucleus. But it can have different atomic weights, depending on the number of neutrons in its nucleus. These are called **isotopes**.

Kelvin scale of temperature   The Kelvin scale of temperature is based on the centigrade scale, but its zero is not the freezing point of water but the absolute zero when atoms have no kinetic energy of motion. On the centigrade scale, that zero lies at -273°C, so the freezing point of water lies at 273°K.

kinetic energy   The energy a body acquires because of its motion.

light year   The distance that light travels in one year: it is equal to about 10 million million kilometres.

local group   The local cluster of galaxies, of which the Milky Way is one of about thirty.

look-back time   The total time that light has taken to reach the Earth from some observed astronomical object.

luminosity   The total rate of light energy output from an astronomical object.

lunar eclipse   An eclipse of the Moon as it passes into the Earth's shadow.

Magellanic clouds   The two small irregular galaxies that are companions of the Milky Way Galaxy, and closest to it. They can be seen only from the Southern Hemisphere.

magnetosphere   The region around a planet that has a magnetic field which influences the physical conditions there. The Earth's magnetosphere is filled with high-energy particles (mainly protons and electrons) that have been emitted by the Sun and become trapped by the Earth's magnetic field. The Jovian magnetosphere is filled by gaseous material ejected from the volcanoes of its closest moon, Io.

main sequence   The sequence of colour and luminosity that stars lie in when they are in that stage of evolution marked by the thermonuclear burning of hydrogen into helium in their inner cores.

mare (plural maria)   A smooth, flat region on the Moon or Mercury created by flows of lava from early volcanic activity.

mass   The amount of matter in a body: in a gravitational field, it causes weight. The term **rest mass** is also used in this book, and this means the mass of a body when it is at rest, that is, stationary. If a body is moving, it has kinetic energy and therefore, via $E = mc^2$, it has additional mass. In normal human experience, this is negligible, but this is not so in astronomy or in particle physics. For example, the mass of a high-energy electron in an accelerator due to its motion can be several thousand times greater than its rest mass.

meteor   The streak of light seen in the night sky when a small interplanetary particle (a meteoroid) burns itself out in the Earth's atmosphere. A meteor shower occurs when the Earth passes through a belt of meteoroids.

meteorite   The remnant of an interplanetary body, which is massive enough to survive the passage through the Earth's atmosphere and reach the ground without being totally evaporated by the friction caused by its high velocity.

micron   A unit of length equal to a millionth of a metre, usually designated "μm".

Milky Way   The band of light in the sky which corresponds to the disk of our own galaxy. It is also commonly used as a name for our galaxy as a whole.

millimetre   A unit of length equal to a thousandth of a metre, usually designated "mm".

molecule   An atomic system formed by the linking of at least two atoms, which results in distinctive chemical properties. Familiar examples are water, composed of two hydrogen and one oxygen atom (H_2O) and carbon dioxide, composed of one carbon and two oxygen atoms (CO_2).

momentum   The mass of a body times its velocity.

nebula (plural nebulae)   An extended astronomical object that appeared to the early telescopic observers to be a nebulous patch. Subsequent investigations showed that the term covers a wide range of very different objects, including clusters of stars, luminous gas clouds, ejected stellar atmospheres and galaxies beyond our own.

neutrino   A neutral subatomic particle of low (possibly zero) rest mass, which is emitted in many nuclear reactions but interacts very weakly with matter making it very difficult to detect.

neutron A subatomic particle with no electrical charge and a mass close to that of the proton (which is positively charged). Together with the proton, it forms the basic components of the nuclei of the elements.

neutron star A star that has collapsed under gravity to a point where the basic subatomic particles have been squeezed together until they are touching, and the overall density is the same as that in the nuclei of atoms. In such a situation, **protons** and **electrons** cannot exist separately and they are forced together to form **neutrons**. To reach this state, a star has to have used up its thermonuclear energy source and be at least 1.4 times the mass of the Sun (which cannot become a neutron star but which will become a **white dwarf**). It will be extremely compact, with a diameter less than that of any major city.

nova (plural novae) A star whose luminosity suddenly increases by a large factor, making it appear as a new star (whence the name). They are now known to be binary stars in which material from the atmosphere of one (a normal star) falls intermittently onto the surface of a compact companion (a **white dwarf**), thereby being heated by the gravitational infall to the point where a thermonuclear reaction is triggered, as in an H-bomb.

Olbers' paradox Developed in the early nineteenth century by the German scientific philosopher Heinrich Olbers, who asked, "Why is the sky dark at night?" At that time it was believed that the Universe was infinite, static and composed of stars, which meant that any line of sight would ultimately reach a star, and the sky should appear as bright as the surface of the Sun. All cosmological theories have had to cope with this so-called paradox.

Oort cloud A very large number of bodies, much smaller than planets which surround the Solar System at a large distance and are believed to be the source of comets.

ozone A molecule of triatomic oxygen (O_3) containing three oxygen atoms. An ozone layer is one component of the Earth's atmosphere and is important to life because it blocks out the energetic ultraviolet radiation from the Sun.

parallax The apparent shift in the position of a star as the Earth moves from one side of its orbit to the other side. A **parsec** is the distance of a star which causes a parallax of one arc second for an Earth movement of one astronomical unit (its distance from the Sun). A parsec is equal to 3.26 light years.

particle physics The study of the subatomic elementary particles that form the basis of matter.

Pauli exclusion principle See **exclusion principle**.

perihelion The point at which any body orbiting the Sun is closest to it.

photon Light has both wave and particle properties. It consists of discrete wave packets (photons) whose energy is equal to Planck's constant (h) times its frequency (v).

photosphere The outer layer of a star or the Sun, from which most of its energy is emitted.

Planckian radiation See **black body**.

Planck's law The expression that describes the distribution of radiation with wavelength emitted by a dense object (black body) at a certain temperature.

planetary nebula The shell of gas surrounding a hot **white dwarf** core, whose ultraviolet emission causes the shell to luminesce and appear planet-like to the early observers. It is an advanced stage of evolution in a low mass star such as the Sun.

plasma A gas that is ionized into positively charged nuclei and free negatively charged electrons. Such a medium is highly conducting and has other properties quite different from those of a neutral gas.

population I stars Stars of the type found in the spiral arms and disk of our galaxy. They are relatively young and contain a higher proportion of heavy elements (C, N, O, etc.). The Sun is a population I star.

population II stars Stars of the type found in the galactic halo and above the disk. They are relatively old, having been formed in the early life of the galaxy, and have a low abundance of heavy elements.

population III stars Stars formed in the very initial birth of the galaxy and therefore believed to be composed entirely of hydrogen and helium, with no heavier elements. So far, none have been observed.

positron The antiparticle of the electron having the same mass but an equal and opposite positive charge.

potential energy Energy that is stored in some way and can be released in certain conditions. Important examples in astronomy are the energy locked in the nuclei of atoms, which can be released in thermonuclear reactions at sufficiently high temperatures (the energy source in stars), and the energy of gravitation, which is released when matter falls in a gravitational field, this is the energy that fuels the exceptional power generation in quasars and X-ray binaries.

precession The slow rotation of the Earth's axis, which precesses like a top with a period of 26000 years; this causes the apparent position of stars to change accordingly. It also causes the signs of the zodiac to move through our seasonally based calendar.

prominences Structures of hot gas above the solar limb, which appear like loops and are constrained by magnetic field configurations.

proton One of the fundamental components of the nuclei of atoms (the other is the **neutron**). It is much heavier than the electron and has an equal but opposite electric charge.

pulsar A rapidly rotating (typically 10–100 times a second) neutron star, from which radiation is emitted in beams along its magnetic poles. Since the magnetic axis is tilted compared with the rotation axis, the beams scan the sky like a lighthouse, and, if the Earth lies in the path, it will see a pulse of radiation every time the beam sweeps by.

quantum mechanics The study of the basic nature and behaviour of matter, caused by the fact that subatomic particles have a wave property as well as a particle property.

quasar An external galaxy whose nature and properties are dominated by an intensely active central core, which often makes it appear like a star (whence the name, a contraction of "quasi-stellar object"). The energy source is a massive central black hole which is fed by surrounding material in the form of gas, dust and stars. The immense power output allows them to be observed at greater distances than any other astronomical object.

radial velocity The component of velocity of any astronomical object in the line of sight of an observer. It is measured by the Doppler effect and can be either towards the observer (blue shift) or away (red shift).

radiation belts Belts of energetically charged particles that surround the Earth and some other planets (see **magnetosphere**).

radioactivity The spontaneous emission of energetic rays from some heavy elements (such as radium) whose nuclei are not completely stable. There are three quite different kinds of ray, designated alpha (helium nuclei), beta (electrons) and gamma (ultra-high energy photons).

radio waves Electromagnetic waves of low frequency and long wavelength (millimetres to kilometres).

red giant A star that has exhausted the hydrogen fuel in its core and has started thermonuclear burning in the shells outside the core; this causes it to expand and, at the same time, to cool. When the Sun becomes a red giant, it will engulf the Earth's orbit.

redshift The shift of a spectral line towards longer wavelengths, caused by the recessional velocity of an object compared with the observer. It is expressed as the wavelength shift divided by the unshifted wavelength and is always denoted by Z. Z reaches infinity only when the velocity of recession reaches that of light.

relativity See **special theory of** and **general theory of**.

rest mass The mass of a body when it is stationary (see **mass**).

rest wavelength The wavelength of some standard frequency when emitted by a body which is stationary.

retrograde motion The apparent motion of the outer planets when they appear to move backwards when viewed from the Earth.

scientific method The discipline imposed on science whereby all assertions, hypotheses or theories must be subject to the test of all relevant information and not just a convenient selection.

Seyfert galaxy A galaxy that shows a spiral structure but which has a very bright nucleus believed to contain a central black hole that generates gravitational power from the infall of surrounding material in the form of gas, dust and stars.

siderial day The true period of the Earth's rotation as measured relative to the stars. It is about 4 minutes shorter than the **solar day** of 24 hours, because this is measured relative to the Sun, and since the Earth undergoes an extra rotation in its revolution around the Sun, there is one additional siderial day in every year.

siderial year The time it takes the Earth to complete one full revolution around the Sun as measured by the stars. It is about 20 minutes longer than a **solar year**, on which our calendar is based because of the slow precession of the Earth's axis, whose inclination determines the seasons.

solar activity The degree of active events on the Sun's surface, such as flares, prominences and sunspots, which vary in intensity on an 11-year cycle.

solar day See **siderial day**.

solar flare See **flare**.

solar prominence See **prominences**.

solar wind The stream of high-energy particles emitted by the Sun into interplanetary space. Its intensity varies with the 11-year solar cycle.

solar year See **siderial year**.

solstice The two occasions during a year when the Sun is at either its maximum elevation (summer solstice) or at its minimum elevation (winter solstice).

special theory of relativity Einstein's theory of motion about how the laws of physics in one body appear to an observer in another body that is moving at a constant velocity with respect to it. This theory unveiled the linking of energy and mass by the relationship $E = mc^2$.

spectroscopy The study of the individual monochromatic emissions of any body, which is the basis of **astrophysics** and which can reveal much about the nature of the body.

sunspot A region on the Sun's surface that is cooler than its surroundings and therefore appears to be dark. It is an area of strong magnetic field.

sunspot cycle The 11-year period over which the number of sunspots seen on the

Sun's surface varies from a minimum to a maximum and back to a minimum.

supergiant   A massive star which has evolved beyond the stage where it is burning hydrogen in its core only, and has become very large and very luminous.

supernova   A final stage in the evolution of a massive star when it has burned the original hydrogen in its core through the successive elements from helium to iron when all nuclear energy has been exhausted. It then undergoes a catastrophic collapse and hurls its outer atmosphere into space at high velocity. Its brightness increases immensely. The supernova that has been studied most extensively is that which occurred in the Large Magellanic Cloud (a small neighbouring irregular galaxy) in 1987 and is designated 1987A.

synchrotron radiation   Electromagnetic radiation emitted when ultra-high energy electrons interact with a magnetic field. It is highly directional and is the cause of the radiation emitted in the pulses of **pulsars**.

temperature   A measure of the internal kinetic energy of the constituents of a medium. See **Kelvin scale of temperature**.

thermodynamics   The study of energy and its changes from one form to another.

thermonuclear   The process by which the energy contained in atomic nuclei can be released at sufficiently high temperatures by nuclear fusion. A prime example is the fusion of hydrogen into helium, the main energy source of the stars.

T Tauri stars   Very young stars which are still in the process of contraction from the interstellar medium.

ultraviolet   The portion of the spectrum that starts at shorter wavelengths than the **visible** and extends down until it reaches the **X-rays**.

uncertainty principle   The principle developed from quantum mechanics by Werner Heisenberg. It states that there is a basic uncertainty in determining the position and momentum of any subatomic particle. For example, an electron orbiting a proton in the lowest level of the hydrogen atom can only be said to be somewhere in that orbit; it cannot be assigned a position beyond that.

variable star   A star whose brightness varies with time. There are many different types, of which the pulsating **cepheid variables** are notable.

vernal equinox   See **equinox**.

visible light   Electromagnetic radiation to which the human eye is sensitive and is seen as the colours red, orange, yellow, green, blue, indigo and violet. It is a narrow band of the spectrum which lies between the **ultraviolet** and the **infrared**.

wavelength   The distance over which any kind of wave completes a full cycle, say from peak through trough and back to peak. In this book it is used almost exclusively for **electromagnetic radiation** from gamma rays to X-rays, through the ultraviolet, visible and infrared regions, to the radio waves.

white dwarf   The final stage of evolution of a low mass star such as the Sun. It follows the **red giant** phase, when its outer atmosphere is ejected and appears as a **planetary nebula**. The core is left to contract and, in so doing, heats up to form a compact white dwarf of extremely high density. Unable to generate any more nuclear energy, it will cool slowly and ultimately become a **black dwarf**.

Wolf–Rayet stars   Very massive, very hot and very luminous stars that are shedding matter in their winds at a very high rate (up to a solar mass in 100000 years). They are evolved objects and the heavy winds reveal the thermonuclear products of carbon, nitrogen and oxygen from the earlier phase of hydrogen burning.

X-rays   Electromagnetic radiation of high frequency and short wavelength, intermediate between the **gamma rays** and the **ultraviolet**.

zenith   That point in the sky which is immediately and vertically above an observer.

zodiac The band in the sky in which the ecliptic plane lies and through which the Sun and planets move. The ancient Babylonians divided it up into 12 equal zones and assigned to each its nearest constellation such as Aries, the Ram, which is now the first sign of the zodiac.

Bibliography

Blacker, C. & M. Loewe (eds) 1975. *Ancient Cosmologies* . London: George Allen & Unwin.

Bronowski, J. 1973. *The Ascent of Man.* London: British Broadcasting Corporation.

Cochrane, C. N. 1944. *Christianity and Classical Culture.* Oxford: Oxford University Press.

Dicks, D. R. 1970. *Early Greek Astronomy to Aristotle.* London: Thames & Hudson.

Farrington, B. 1936. *Science in Antiquity.* Oxford: Oxford University Press.

Finley, M. J. 1963. *The Ancient Greeks.* London: Penguin.

Friedman, H. 1990. *The Astronomer's Universe.* New York: Norton.

Gingerich, O. 1992. *The Great Copernicus Chase and Other Adventures in Astronomical History.* Cambridge, Massachusetts: Sky Publishing; and Cambridge: Cambridge University Press.

Glanville, S. R. K. (ed.) 1942. *The Legacy of Egypt.* Oxford: Oxford University Press.

Koestler, A. 1959. *The Sleep Walkers.* London: Penguin.

Kuhn, K. F. 1991. *In Quest of the Universe.* St Paul, Minnesota: West.

Manuel, F. E. 1980. *A Portrait of Isaac Newton.* London: Frederick Muller.

Moet, A. 1927. *The Nile and Egyptian Civilisation.* London: Kegan Paul, French, Trubner.

Murray, M. A. 1949. *The Splendour that was Egypt.* London: Sidgwick & Jackson.

Pais, A. 1982. *Subtle is the Lord. . . the Science and the Life of Albert Einstein.* Oxford: Oxford University Press.

— 1986. *Inward Bound.* Oxford: Oxford University Press.

Pasachoff, J. M. 1991. *Astronomy: from the Earth to the Universe*, 4th edn. Philadelphia: Saunders.

Peebles, P. J. E. 1993. *Principles of Physical Cosmology.* Princeton, New Jersey: Princeton University Press.

Riordan, M. & D. N. Schramm 1991. *The Shadows of Creation.* San Francisco: W. H. Freeman.

Shipman, H. L. 1982. *Black Holes, Quasars, the Universe.* Boston: Houghton Mifflin.

Shu, F. H. 1982. *The Physical Universe.* Mill Valley, California: University Science Books.

Snow, C. P. 1981. *The Physicists.* London: Morrison & Gibb.

Snow, T. P. 1991. *The Dynamic Universe*, 4th edn. St Paul, Minnesota: West.

Zeilik, M. & E. V. P. Smith 1987. *Introductory Astronomy and Astrophysics.* Philadelphia: Saunders.

Loeb Classical Library: *Aristotle*
Volume v: *Physics*, translated by P. H. Wickstead & F. M. Cornford, 1934; Volume vi: *On the Heavens*, translated by W. K. C. Guthrie, 1939; and Volume xv: *Problems*, translated by W. S. Hett. 1953. Cambridge, Massachusetts: Harvard University Press; and London: William Heinemann.

In addition to the above, the writer has drawn on many learned publications, including *International Astronomical Union Symposia* (Dordrecht: D. Reidel), *Annual Review of Astronomy and Astrophysics* (Annual Reviews Inc.), *Quarterly Journal of the Royal Astronomical Society* (Oxford: Blackwell Science), and *Nature, the International Weekly Journal of Science* (London: Macmillan).

Index

T - #0108 - 111024 - C84 - 229/152/16 - PB - 9780367400880 - Gloss Lamination